Secrets of the Still

A Zesty History and How-To for Making Spirits, Fragrances, Curables, Gasohol and Other Products of the Stillroom

Grace Firth

EPM Publications, Inc.
McLean, Virginia

EPM Publications, Inc. 1003 Turkey Run Road, McLean, Virginia 22101
Printed in the United States of America

Book design and illustrations by Tom Huestis

Library of Congress Cataloging in Publication Data

Firth, Grace.
Secrets of the Still.

Bibliography: p.
Includes index.
1. Distillation–Amateurs' manuals. 2. Fermentation
–Amateurs' manuals. I. Title.
TP156.D5F5 1983 660.2′8425 83-16369
ISBN 0-914440-66-7

PREFACE

A certain caterpillar is said to weave simultaneously a number of different silks to form a cocoon. So judiciously spun is the structure that no two threads become entangled. Would that I possessed that talent in weaving the threads of distillation through the various geographic, materials and time webs involved in man's development of the still.

In an attempt to hold the strands apart, the areas of flavorings, fragrances, medicinals, potable spirits and fuel alcohol have been separated even though in reality each was interwoven with the others' distillation loops.

Sketches of history, some necessarily speculative, are included in the hope that they will furnish a record of the place and function that spices, spirits and the search for curables played in man's crisscrossing the planet Earth.

Though few chemists would employ cups or spoonsful in formulas, for consistency and ease of understanding measurements are given in U.S. customary and metric measures.

SECRETS OF THE STILL was written as an introduction to the fascinating subject of distillation and as an informative and entertaining overview of stills. Although extensive efforts have been made to validate the factual information used in this book, early-day beliefs were sometimes erroneous. Neither the author nor the publisher makes any warranty, expressed or implied, or assumes any legal liability or responsibility for the accuracy, completeness or usefulness of the information, the apparatus or processes presented.

Obviously, this book offers only a sample of the numerous types of stills and distillation processes man has devised. In addition, it barely touches the endless variety of substances that man has distilled into fragrances, healing potions and brandies. Like a box of assorted chocolates, this book is a sampler; it was written to tease the inquiring mind and yet satisfy the sweet tooth of an armchair explorer.

It would not be possible for me to list all of the people who have answered my questions about distillation, but it is a pleasure to acknowledge the assistance of librarians of the Library of Congress; Crimora Waite and her staff at the Culpeper Town and County Library, Culpeper, Virginia; and the library workers of the Prince Frederick, Maryland, library. Both Ruth Smith and Tom Birkel read the manuscript and made valuable suggestions; Ruth lent her expertise in medicinal herbs and Tom, my technical consultant, constructed and taught me how to use my steam still. Lise Metzger edited with a clear eye for organization and my daughter, Penny Kondratieff, explained to me some complexities of distillation. To all of these people and to my patient publisher I give my thanks. But above all others who were with me in spirit and in fact throughout this wonderfully provocative and satisfying endeavor, was Lewis, my husband, and I thank him with all my heart.

Heartfelt thanks, tangled threads and all, I pray that I have opened new windows in the reader's mind, and that I have communicated my own feelings of real respect for those who have written more learned accounts of the art of distillation.

Grace Firth
April 1983

CONTENTS

This book is for Lewis.

1

CHAPTER

Keys to the Stillroom

I made my very first still in Birds Nest Cove near Sitka, Alaska. We were stranded and there was no water except the sea. Our adventure had begun like many spur-of-the-moment Sunday afternoon ideas that sound good at the dinner table. "Let's go!" I was single, it was a sparkling day in mid September, and with our spirits rocketing, four of us fired off for the island of Birds Nest. Though the tide was high and we bumped a few times as we bulleted over the rocky breach into the cove, not one of us realized that after the equinox tide receded there was no way out of that watery graveyard of boulders. At first we were enchanted by the place; mountains cut the deep blue skyline, black shore woods whispered native legends, and looking back to the North Pacific entrance to our paradise, we saw the golden sun play on the water frolicking among the rocks.

Suddenly, as of one mind, we understood that until the next extreme high tide, which might not occur until October, ours was a voyage of no return. With no provisions, charts or wit to foresee the seriousness of our plight, we watched the retreating tide expose a rock pathway to the sea, and try as we did, we could not maneuver our boat out of the cove. At darkness we huddled around a beach fire and passed the time crushing mosquitoes. It was a long night.

The following day we turned to search for roots, greens, anything that moved, and—most important—air's first cousin, fresh water; but there was no water, and by the second day our thirst became overwhelming. Remembering a neighbor's flower-water cone still and the coppery contraption in grandma's cupboard above her kitchen sink (a place into which I had been forbidden to peek) I scrubbed clean our coffee-can bailer and gas funnel while two others paddled into the lagoon for the least turbid water. The boat owner, Mike, removed the outboard motor top plate, which had a dip and donut hole in the center.

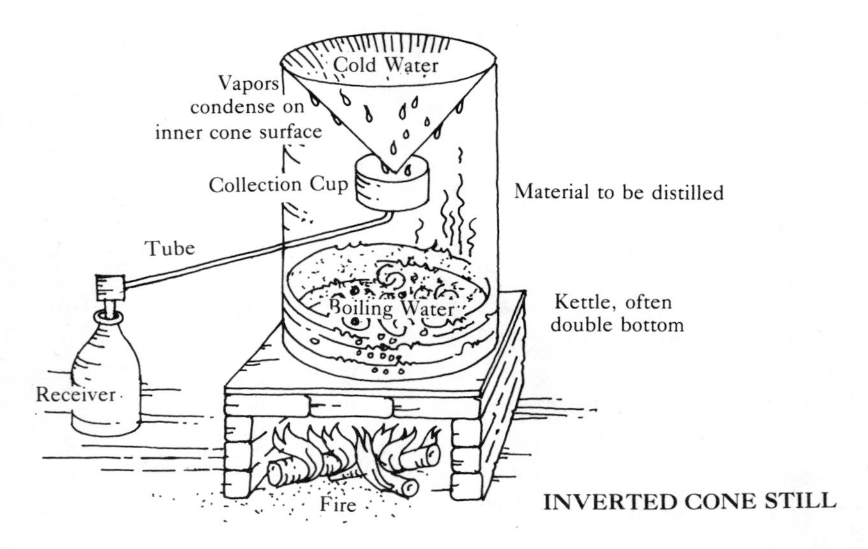

INVERTED CONE STILL

First we fitted flat rocks into the bottom of the galvanized bait bucket. Placing the coffee can on the rocks, we weighted it so that it would not tip. Next we fitted the funnel into the motor plate hole so that its spout continued the plate dip and was centered on the coffee can below. After stuffing moss into all cracks, we built up our fire and began to boil sea water in the bucket.

She worked! Beginning with drips, we soon heard steam condensation running off the top plate, onto the funnel, down into the coffee can. Unfortunately, it tasted like oily sea water. But a day later our nifty little still perked pure flat H_2O. That was also the day it rained. When, after six days of isolation, a search plane spotted us, we had mixed feelings about returning to civilization; sharing and overcoming a hardship binds people in a way that is akin to love.

I learned several lessons from our Birds Nest frolic: Sunday afternoon inspirations are not to be trusted, and never boil your still dry unless you enjoy burned squid juice. I also learned to make pure, colorless, odorless, tasteless, distilled water. Rain is nature's distilled water, and if it does not fall through dirty air, it is pure. To remove the flat taste, mix in air by shaking it in a container.

In today's world the rogue, artist, gentleman and distiller belong to no class. In addition to classlessness, the distiller of spirits, like a purveyor of illicit love, suffers the evil eye of moral condemnation. Busybodies gossip, authorities put the tax hex on him, his privacy is shredded by peeping pedestrians, and deputies sniff after his fumes. The tool of his trade, his still, that simple innocent, is an outlaw. In the United States, the possession of a potable spirits still is not sanctioned by the official rules of the legal game. A thirsty man may distill water, a cook may distill herbs, a chemist may distill wood for a law-abiding turpentine, a beautician may distill roses, a doctor may distill witch hazel, a farmer may distill grain for legal fuel, a citizen may distill all of the above, but a moonshiner is considered knavish, dishonest, perfidious, venal, fraudulent, corrupt and just plain foolhardy if he distills his own hooch. Just owning a still for making spiritus vin can land you in the slammer.

Years ago distillation was a socially accepted act. Parsons did it, spinsters were good at it, women and men coveted their skills in the art. Like milking a cow, when the time came to take off a little juice, the operation was set up and herb flavorings, medicines, a drop of rose water or brandy was squeezed out. Some distillers bartered their products; four bushels of grain could be traded for a jug of spiritus vin. In time, governments caught on that they had a real patsy. People loved spirits. Excesses led to badness. Badness was bad. Badness should be punished. Pocketbook pinching was a condoned punishment for badness. So governments began lining their pocketbooks with pinched lucre; politics and war took lots of lucre.

However, because some spirits distillers believed that the tax on liquor was disproportionate to the tax on other enterprises, they refused to pay the tax and moonshine operations sprang up.

Traditionally, the distillation of beverages has been only one part of a still's employment. During the Middle Ages, civilization reeked of foul odors, food wallowed in monotony and home remedies healed everything from carbuncles to kidney warts. Simple distillation of fragrant and flavorful herbs gained popularity.

These same times saw the stillroom develop out of need. Initially a stillroom was a hiding place for food. Later it evolved into a heatless nook near the kitchen where homemade curables, caches of wines and aromatic waters were tucked away with fermented cheeses, krauts and a sausage or two. As the Middle Ages waned and marauders preyed less frequently upon their brothers, the stillroom matured into a legitimate appendage to the home: a pantry where a wife fermented foods, the brewmeister watched over his wort, the stillman tended his still and the sneaky nipped a bit of schnapps.

A STILLROOM

My granduncle's neighbor had an old-timey stillroom built underneath the porch steps. The man was known as a horse doctor; people said that he could hack off teeth, heal the staggers or pull a colt with his home-stilled colic medicine. As a child I peered down through the grating on the basement window of his stillroom, and when those musty odors rose through the weird copper contraptions and hit my nostrils, thoughts of tawny men brandishing curved knives flooded my brain. Though long on fantasy, I was short on courage, and after one eyeful I usually retreated at a run to my uncle's chicken yard, where I played with a crazy hen who regularly knocked herself flat by challenging the grinding wheel when I sat and pumped it. Uncle

Otto's neighbor distilled anything that "creeps, seeps or crawls" into his stillroom, people told me. I believed them. I never went down inside.

A still, the nickname for "distill," generally refers to an apparatus used to heat a substance, vaporize the volatile constituents of a material and condense the vapor into a liquid. Distillation is a natural process. Thanks to distillation, rain leaps to earth with its gift of hope.

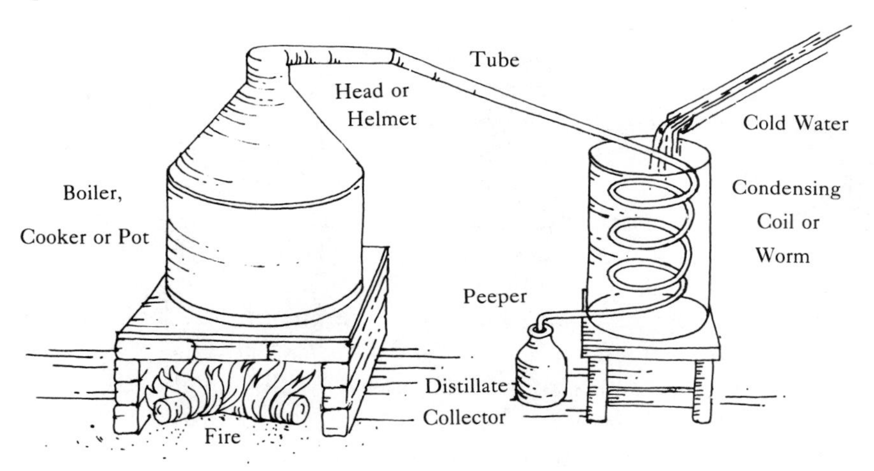

PARTS OF A SIMPLE STILL

Still designs have traditionally consisted of three parts. The boiler, also called the pot, can, cooker or kettle, is used to boil the raw material; the head, helmet or bonnet is a smaller compartment that sits atop the boiler to gather and compress vapors from the boiler and direct them into the condenser. The condenser, often called the worm, is a copper coil or straight pipe immersed in water whose duty is to take the compressed vapor, increase its density toward its saturation point, then cool the vapor in order to turn it back into a liquid.

Different substances vaporize, or boil, at different temperatures. For example, alcohol boils before water; a mixture of the two can be heated, and alcohol vapors can be collected and condensed back into liquid before much of the water turns to steam. Thus alcohol and water can be separated.

Fermentation and Distillation

In studying stills it becomes apparent that distillation has evolved hand and glove with fermentation. Fermented milk to create the cheese products of the primitive Indus valley civilizations, fermented bean curd and pickled fish of Asia, fermented breads of ancient Egypt, all originated several thousand years B.C. Vinegary foods, soy sauce and hard sausage, all of which undergo fermentation by microbes, were eaten by people living hundreds of years before Christ, and archeologists attest that fermented beers and wines helped wash down oodles of ancient meals. Beermaking has been documented as existing in 1750 B.C., and vintners were experts at squeezing a grape long before then.

"What is the relationship between fermentation and distillation?" a modern young woman asked me. Because crocks of ferments (pickles and herring as well as brews) had been a part of my young life, I had assumed that everyone knew about ferments and I looked closely into the trim girl's face. There was no trace of sham.

Yeast fermentation is preliminary to the distillation of potable spirits and fuel alcohol, I told her. Yeasts that ferment a sugary liquid change the sugars into alcohol. By distilling the liquid, the alcohol can be separated from it. Simply stated, fermentation is a chemical changing process, distillation is a separating process.

The process of fermenting liquids is similar for both wine and beer. Fruit juice, or water and malted grain (often mixed with sugar and flavoring), are charged with yeast and allowed to brew, or "work," in a warm place. The yeasts eat the sugar and give off alcohol and carbon dioxide, which is the bubbles. When most of the sugar is eaten, the brewing slows and the beverage must be sealed to exclude air; thus the wine or beer is preserved from being turned sour by a second army of microbes. An alternative to sealing beverages is distilling.

The process of distilling wine to form brandy, or grain beers to form spirits, entails heating the liquid, converting it into vapor, condensing the vapor in a cooling tube and catching the condensate. The process is the same as distilling water except that the heat is kept lower than that of boiling water because the object is to collect the alcohol, and alcohol vaporizes before water. The trick in making distilled beverages is to collect that which runs through the still first. Both wine and still beer (beer made specifically for distillation) contain about one part alcohol to about ten parts water, so distillers rarely run more than a quarter of the fermented brew before stopping the process. The distillate that is collected may be distilled a second or third time to concentrate the alcohol.

I went on to explain that microorganisms have been working their little tookies off for man for a long time, that microbes need food (such as sugar), warmth and moisture and that some ferments need air to thrive. Most ferments enjoy a near-neutral pH environment (pH is a measure of the relative acidity of a solution). Remove any one of the basic life needs and most microbes cannot survive.

Fermentation, the changing of organic compounds, is achieved by means of bacteria and fungi as well as by yeasts. If food, temperature, moisture and air systems are go, bacteria multiply like crazy. They eat their host's sugars and quickly give off an acid that inhibits growth of acid-intolerant microbes. The high acid condition present in cheeses or pickles discourages the growth of "bad" bacteria. Different parts of the world have encouraged the growth of their own "good" bacteria, and by their very numbers these microbes impede the growth of undesirable bacteria that may cause spoilage. Examples of this are cheese factories and sausage-making shops that are so infested with "good" bacteria that "bad" microbes cannot get their fangs in the food.

Various Uses of the Still

In addition to the employment of fermentation for flavoring, making food tasty by using distilled herbs and spices has a long history. Plants have been distilled for their essence (technically known as volatile plant oils) since witches and warlocks wearied themselves over caldrons and the Arabians of the 1100s extracted the "spirit" from plants. Unfortunately, distilling during this period became sidetracked. After enjoying unrivaled success at separating the volatile oils from plants (such as distilling oil of rose from roses), Arabian scientists tried to distill the "spirit" from all substances. Their search evolved into astrology, the belief that "spirits" of heavenly bodies shape human lives. Interest in the distillation of plant materials, whether exploring flavorings, fragrances or medicine, was eclipsed by fascination with the zodiac.

I knew a warlock once, a tall redhead whose teeth vanished from his mouth when his troller hit a reef and he went head first through the wheel. He had been climbing from the bilge when his craft grounded, and after his rescue he called himself Warlock, Champion of the Black Arts. He never fished again. Instead he painted his name over an abandoned warehouse office that he occupied in Seward, Alaska. He entertained visions and vibrations, drew horoscopes and distilled odoriferous materials that steamed his window and polluted the lower end of town.

It is a shame that the still and its uses have been blighted by association with omens and signs.

Distilling plant parts, extracting their flavor for use in sauces, bakery goods and beverages, is an honorable pursuit. Plant material, such as basil leaves, is placed on a rack in the boiler of a still, water is added, the still head is fastened, the condenser is checked to see that its tube is submerged in cold water and a fire is lit under the still. Within moments a pungent vapor starts rising and seconds later the first drips sputter out of the worm. Bam! A full shot! But a distiller must be nimble. Most essential oils volatize quickly, and unless you wish a watery distillate, the still must be quickly closed down and the essence corked.

Distilled fragrances are equally rewarding and fun to extract. Handsful of flowers are tucked into a steam still (a still in which the water is boiled in a separate vessel and its steam is piped to the bottom of the flower container). Adding a little water to the still keeps the petals lubricated, and when the injected steam bubbles up through the fragrant material, its heat causes the volatile oil sacs to rupture. Bam! Fragrant steam rushes into the condenser. Cold temperatures of the condenser squeeze vapors back into a liquid and scented waters exit through the tube.

During early days, fragrant waters were the backbone of homemade cosmetics. Physicians of yore also distilled plant substances to make curables. In some locations physicians were given the sole right to own and operate a still. They not only explored the healing qualities of specific herb distillates but also pinpointed the idea that exposure to air, heat and fermentation changed many plant substances. In some cases these changes created new properties, and scientists began compounding "virtuous" qualities of herbs. Rose oil and chamomile water were combined to form a sedative. Warmed flax seed (linseed) oil was compounded with bramble berry leaf "tea" to relieve gas and colic. The ability of sage water to induce menstrual flow was coupled with the quieting qualities of distilled rosemary.

Stills come in all sizes. Some are as small as coffee mugs, others are as big as box cars. Though stills are usually made of copper, stainless steel or glass, I have seen them made of tin, earthen materials with wooden tubing, and galvanized metal. I saw one still made from a church incense burner and moose antlers.

The Strangest Still

In February 1946, three friends and I skied from the center of Alaska's Kenai Peninsula to Ninilchick near Cook Inlet. There were no roads. Our party consisted of a couple of former nightclub adagio dancers called the Shadrows, a pilot from New York named Goldman, and me. On one thing we all agreed: We would be experienced cross-country skiers by the time we completed our eight-day trek across the rolling, stunted pine lowland. In addition, Goldy and I discovered that we would be experienced mediators in family arguments.

On the morning we left, the sky hung like a shroud, there was no wind, the air was like fresh milk. We should have taken warning from the midwinter thaw because before we had followed the river westward a half a day we were drenched with sweat, and despite repeated ski waxing, hunks of balled ice clung beneath each ski. Two miles out, the Shadrows, who had been compatible colleagues at the Indian boarding school where I taught, started fighting over which of them had forgotten their extra socks. By four miles, Goldy had worn a blister on his heel, and before making our first camp, I had fallen through the ice and was soaked to the belly button. Each morning thereafter the four of us woke to find our sleeping bags in the bottom of a hole of melted snow and our fire drowned. Each day was a repeat of arduous trials with balled ice, patches of devilish briars, lost bearings and the everlasting bickering between the dancers. At one point, the lady, all fire and passion, declared that she was leaving the troop; she turned back and her partner plowed ahead. After about an hour we retraced our steps and found our companion sitting on a stump eating chocolate. She nearly strangled us all with embraces. A moose had frightened her and she vowed that she would not leave us again.

As we neared the coast we saw a plume of black smoke reach into the white sky and layer itself across the horizon. We ran across dog trails, garbage dumps, cut trees and finally we approached an apparently abandoned log chapel squatting on a bluff a distance from the village.

Inside we found a poorly clad, swarthy young woman. Her black hair was tied back with a yellow bandana and her eyes, though sober, wore wrinkled remnants of smiles. The girl was making tea, she said.

Walking to the low steps upon which a contraption hummed, I could not help noticing that her teapot was a still. Our eyes met, and lowering hers, she said, "My husband . . . that is . . . the beer, it will not wait. He has been taken to the hospital." She darted a swift glance at me. Her face was simple, round, agreeable and cast with a melancholy that gave it strength. "He hurt his shoulder. The men of the village burned tires; it is the signal to call a plane." Her words were clipped in the native way and her voice was steeped with sadness. Bright tears suddenly swam in her black eyes. I would have liked to have comforted her but the Shadrows were intent upon skiing over to Ninilchick and signaling a plane to pick them up. I knew that Goldy's feet were rubbed raw. So taking another look at the still steaming boisterously on the altar steps, I saw that it had been fashioned from an ancient brass incenser with an elaborate hollow handle, which had been extended by means of a hollowed-out moose antler to form a condensing tube. Standing by my shoulder the girl murmured, "You cannot waste good ferment, it is the mother of life. Like the sun it gives and destroys life." She stepped to her still saying, "With ferment you capture ferment. Forgive me. . . ."

The plane that landed on the narrow sand strip below the bluff could carry only three passengers, and while waiting for it to return for me, I inquired of the storekeeper about the young wife.

"She was probably running off Sim's still beer. He makes his own likker. She's from one of them Indian schools in the states," a villager told me. "Sim got hisself bumped up when he jumped off the bluff."

Sim, Simeon Mitchell, was the son of a precise Seattle typesetter and a half-blood Indian wife who smothered her only child with love. "Sim growed up as if under a mushroom," the man informed me. At his parents' death Simeon sought the bottle, and from that day an odor of liquor surrounded him. Known as a good-hearted tramp, who, when he had money, headed for the closest bar and stood treat to all comers, Simeon was also a compulsive gambler. He took bets on anything: fleas on a dog's ear, salmon in a gill net, a razor clam's speed, raisins in a pudding.

Three nights before our arrival at Ninilchick he had bet his companions that he could jump off the Inlet bluff and not ruin himself. The cliff's bottom, a distance of about 40 feet, could not be seen from its lip. After collecting $17 and a bottle of booze, "Sim sprung like a frog," the man told me. The whole village had turned out to watch and for a long moment after he had leaped, no one spoke. They were all horrified, the man said.

After a time they heard a hollow voice rise out of the cleft. "I'm O.K., but the way down was long. It dried out the part I like best, my whistle. Toss down a flask."

Later a rescue line was lowered because an incoming tide had prevented Sim from walking to the village. In the darkness the rope had fouled, and though men had worked furiously throughout the night, the snag could not be freed. A second rope of sufficient strength could not be found, and Simeon had hung by his armpits pleading for the still in heaven to quench his thirst. Like the scream of a hawk, Simeon's tirade against his own spendthrift life, his own worthlessness, his pledge for amendment, was almost too shrill to be heard. But his wife, kneeling on the black brink, heard his vow to stop drinking and she clung to his words.

When Simeon was finally rescued the following dawn, his first cry was for whiskey.

The wife with her still in the derelict chapel was not yet at peace when I met her. She was inwardly raging against herself, but she could not waste ferment. With ferment you capture ferment, she had said.

SECRETS OF THE STILL was written to guide the reader along the beguiling paths of fermenting food and beverages. It was written to illuminate ways of creating flavorful distillates. The book peeks into the tangles of capturing fragrances and making simple cosmetics, and it takes a questioning look into the realm of curables. SECRETS warns the reader about the hazards of making illicit whiskey as it looks over the horizon to the use of renewable energy sources such as fuel alcohol. SECRETS OF THE STILL is aimed at making man aware of the need for respect and repair of the damaged earth. It is a book about the natural process of change—fermentation—and about distillation; it is an excursion into the secrets of stills.

2

CHAPTER

Stills in History

There is a pleasure in studying stills. People tell tender tales about shadowy memories of the past.

Doubler Jones was a moonshiner. He had a lean, red face, a rubbery nose, an old felt hat and a throat that seemed to be lined with grit. "Stir that mash, son," he yelled in his gravelly voice. "Gotta cool her down 'afore we can spark her with yeast."

A lad, no more than ten years old, reached, arms extended over his head, to heave backward and forward on a wooden paddle protruding from a huge fermenting tank that was nearly full of soupy, cooked grain. He knew that when the grain cooled to body heat his father would add malted barley and sugar as well as the yeast. He had been told that malted barley would help turn cooked grains to sugar and that the yeast would turn the sugar into alcohol. The boy knew that they were making illicit alcohol for sale to a local "runner." (The runner then would carry it to a "traffiker," who would peddle it to a "buyer," who distributed the "likker" to bootleggers in the city.)

Even as he stirred, stepping forward and back with each sweep of the hook-armed paddle, the Appalachian youth understood that temperature was important in the fermenting of beer. At the moment, the sloppy cracked grain was hot from its cooking, but after it had been cooled and charged, it would be covered with quilts and left alone to ferment for about five days. At that time he would help his father bucket out the milky still beer and carry it to their still, located on the other side of the spring run-off.

Now, stirring the heavy liquid, the boy looked across the shallow ravine where the stream meandered through the brushy woods. He watched the still hand bend to stoke the fire, which had been built in a depression in the ground. He saw the blackened cooker part of the still sitting squarely over the fire. He could hear it sing as it began to boil. And as pressure from the cooking beer forced vapors to squeeze into its copper still head and through its homemade beak that extended a yard's distance out from the boiler, he heard the click and crackle of constrained steam. He knew that the still head's beak poked into a pipe that was elbowed into a small wooden keg propped in a saw horse a short space from the hot boiler. The thumper keg putt-putted like a motorcycle as each burst of steam from the still spit vapor into it. "Boy, if I had me a bike! Maybe the old man will buy me one when he sells this batch. I could be a lookout down at the crossroads. If I saw anyone coming up the road I could zoom up here 'n warn him."

The boy, a gaunt, timid youngster with skin, eyes and hair that appeared to have had their color washed out, paused, a flicker of desire softened his face.

His father, who was standing beside the thump keg across the ravine, did not look up. He was listening to the steam from the still as it spewed into the keg. Part of the watery vapors were condensing in the keg; he could hear liquid drip into the residue in the keg bottom. He gingerly felt the thump keg's copper exit pipe to see if it was hot enough to be carrying alcohol vapors into the big coiled condenser tube that was immersed in a nearby barrel of cold water. Satisfied that some of the water in the distillate had dropped into the thump keg, and that alcohol vapors were continuing on their way to the condenser worm, he looked up. "Your mash getting cool enough to touch?" he called to his son. Reaching and testing the soupy grain with his hand, the youngster nodded and asked how much sugar and malt he should add to the mash. He tried to force his voice to be low and rough sounding like his father's.

"Hold it over there," the older man called back. "I want to test this stuff first, see if she's running watery. I'll be right over to set-by your mash, get her ready to ferment next week's run." As he yelled, he bent and put his forefinger and thumb under the clear dribble that was sputtering from the end of the copper condensing coil, which emerged close to the bottom of the water-filled barrel. Feeling the slick, almost cool liquid that dripped into the collector, he turned to the man stoking the still boiler fire. "Let her run another five minutes, then close her down. I'll be over here with the boy. This is our last run, I'll help you clean up soon as I set the new beer to brewing. Be back in a minute, we'll clean up, then double."

Doubler Jones earned his name because he always distilled his whiskey twice. He was a doubler. He filtered his first-run whiskey through charcoal, then ran it through the still a second time to make his product crystal clear; a first-class shine that would kill a roach on contact.

The man who told me about his youthful experiences distilling illicit whiskey in the mountains of Virginia was the son of Doubler Jones. Today he is a respected executive of a chemical firm. He learned about running a business from his early moonshine training, he said. He learned about buying supplies on the cuff and the agony of paying back debts, about producing a good product, working hard and hearing the jingle of cash. And though he had not helped make moonshine since he was "a sprout," he had found out about distillation, he asserted with a nostalgic smile.

"He found me," his wife piped as she reached to press his hand. "A good woman is the distillate of a man's hopes and dreams," she added.

She was right, but since then I have learned that there is a lot more to the study of distillates and stills than making moonshine.

Ancient Backgrounds of Distillation

In examining the forces that gave rise to the advent of the still, it seems that man's quest to preserve plant fragrances inspired primitive peoples, East and West, to play around with extraction. They wished to save plant scents for use the year round, and they wished to trade their products. The earliest known "distilling" device was an earthen kettle filled with boiling cedar chips or other fragrant material. Wool was layered on sticks across the top of the open kettle and oils from the boiling substance rose with the steam to saturate the wool. The "distilled" watery oil was later squeezed out.

Spices include aromatic herbs, roots, bark, buds, stems, seeds and vegetable substances, and ancient man considered them precious gifts. The oldest artifacts pertaining to spices lead backward to Asia, where people used them as articles of trade. About 3000 B.C. the Indus River

Valley became a spice center that promoted overland commerce between India, the Far East and the Middle East. As growing populations demanded more spice, Indian merchants had to seek cinnamon bark from Ceylon, black pepper from Borneo, cloves from Java; these were shipped up the river and overland by camel caravan to Baghdad and points west.

OIL BEARING VAPORS COLLECTED IN WOOL

Early-day Egyptians, knowing a good smell when they sniffed one, grabbed onto spice to make aromatic oils, which were probably a mixture of spices and heated fats. We know how butter absorbs icebox odors; with repeated exposures, oils become saturated with an essence. The ancients of the Nile expressed oil from flax and radish seeds, nuts and olives for use as food, medicine, illumination and body lubricants.

Fragrant oils also were employed by Egyptians in embalming the dead. They associated aromatic scents with purification. By the application of scented oils and the burning of fragrant wood, a person's spirit was honored, and it was believed the spirit so honored would gain merit in the sight of their gods. Thus spices took on symbolic meanings and were integrated into religious rituals.

In addition, dry climates encouraged Middle Easterners to anoint themselves with perfumed fats. I have had my troubles with sun oil deposits on clothing and I feel sympathy for Egyptian housekeepers, because an oil rub was as popular in those days as today's shower bath.

The distillation of spirited beverages probably originated with early Egyptian beer. A traditional chore of women, beer making most likely evolved from bread making, during which barley was sprouted, roasted, soaked, made into a dough with wheat before being baked. Sometime between the soaking and dough steps, a batch probably fermented, the little lady tasted the brew, licked her chops and began producing beer. Beer was sold by women from their homes, and it was doled out by pyramid "union leaders." Slaves received about three cups a day and the foreman about eight cups; throughout history foremen have been experts at bending an elbow.

Alcohol content of the ancient brew was said to be between 8% and 10%, which, with a little ingenuity, could be encouraged to change into something stronger. Egyptians were up to the challenge. They developed a box still. Placing a flat bowl of beer in a tight box with a slanted glass lid, they set the contraption in the sun. Desert heat vaporized the beer, vapors rose to collect on the inside of the cool glass, and as more moisture collected, droplets formed and ran down inside the glass into a trough, which by the end of a hot day contained some pretty potent stuff. Problems arose when an ordinary

AN EGYPTIAN BOX SOLAR STILL

person was caught with a still, however, because then, as now, the distilling of spirits was restricted to high priests and nobles of the art.

The distillers of spirits in ancient Egypt insisted upon a closed shop. In addition to coercion, such as stuffing an illegal distiller into his own box still, the princes of distillation created an aura of mysticism about their art. In order to keep out the curious, they devised rituals crediting their skills to the supernatural. These mysteries that surrounded distillation were later transferred to other Arabian countries, and they became imbedded in the foundation of astrology.

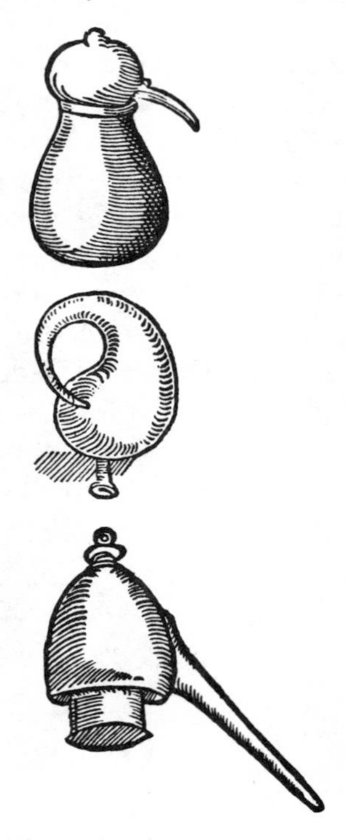

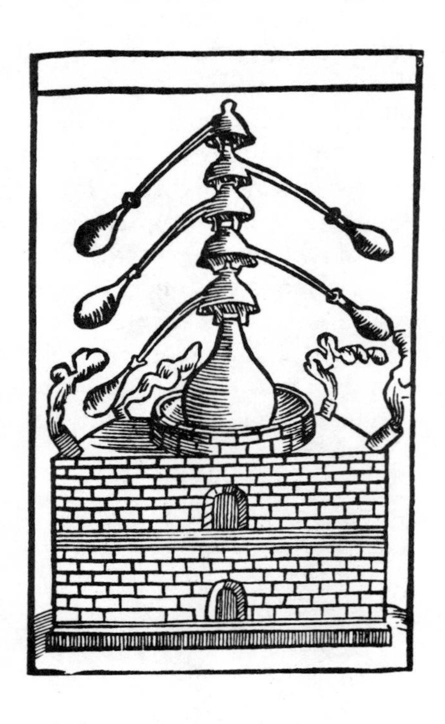

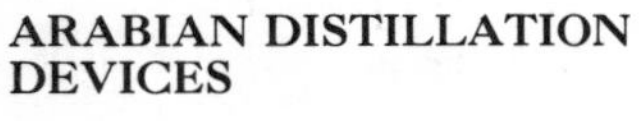

ARABIAN DISTILLATION DEVICES

ARABIAN STILL FOR SEPARATING PLANT "SPIRITS"

Arabian peoples, by virtue of their access to the Indian Ocean, built ships that challenged overland spice routes, and in time they dominated the early spice trade. With Phoenician seamen at the helm, the Arabians founded companies that spread their language, merchandise and culture along the south sea coasts and up and down the Mediterranean. Scholars borrowed Egyptian techniques of distillation and improved them. Though representations of these earliest stills do not exist, it is known that Arabians were familiar with furnaces and that they were obsessed with the alembic: a container with a head and beak. Some alembics boiled plant substances and circulated the steam back into the alembic. In time, the scientists devised alembics in which the beak protruding from the head was extended to form a tube to carry the boiling spice vapors through containers of cold water. Thus vapors condensed into a liquid and the result was a watery distillate.

The integration of volatile plant substances in a liquid was thought to hold special powers. When they boiled plant parts in their alembics, early Arabian scientists believed that they were extracting the spirit from the plants. They regarded their activities as part of a search for an elixir that would maintain mankind's health and prolong life. This search for an elixir fostered the development of distillation.

TWIN ALEMBICS

THE BEAK OF AN ARABIAN ALEMBIC COOLED BY WATER

Alembics were used to purify the "spirits" of plants

Middle Eastern medicine, the art of healing with aromatic waters and unguents made from watery distilled spirits of plants, found pay dirt in the alembic contraptions, and their use came to be under the jus divinum of physicians. Men of medicine cleansed sinuses with vapors of ginger; fiery cinnamon oil distillates were applied to sores as a caustic; juniper water baths and applications of spiced oil were prescribed for syphilis; and clove water was taken for stomach upset.

For hundreds of years before the time of Christ, the demand for fragrant spices increased in all Middle Eastern countries. The wealthy burned spices on smoldering fires, women pomaded their hair with scented fats, men luxuriated in fragrant bath oils, and aromatic herbs were strewn on floors and among clothing.

Seeing the ready-made market and the riches in spices being siphoned off by Arabian port cities, the Persians rehabilitated camel highways to the Orient and put their fingers in the spice jar. Later they introduced their own product for trade, Persian Rose Water. Through its promotion, the Persians were able to spread beliefs that they were superior people, and they conquered the land from Egypt to the Aegean Sea. The distilling of rose petals utilized curiously shaped alembics and more curious heating techniques. Some vessels were submerged in hot springs, in warm ashes, in fermenting substances. Some were heated by burying the alembic in piles of manure. Finally a convex helmet still was constructed for the distillation of rose petals. The alembic boiler fitted down into a brick furnace; on its top a cone-shaped helmet that had a groove around the inside of its base caught the condensing vapors of boiling rose petals, and the groove directed the condensate into an outlet tube.

After Alexander the Great brought Greek domination over the Persian empire (4th century B.C.) the goodness of spicy distillates was spread throughout the classical world. Though few Egyptian and Arabian records remained, their distilling techniques, together with knowledge about the natural sciences, were ferreted out and recorded by the Greeks.

In expanding the understanding of distillation, the Greeks organized their system of simples (as single plant remedies were called), and thus they plowed the ground for Dioscorides' benchmark work MATERIA MEDICA. This medical reference, published during the first century A.D., was regarded as an authority until the time of Luther in the 1500s.

Under Greek control, Persia and the Arabian kingdoms were allowed to continue their caravan and spice-shipping trade with the Orient. There was rivalry, but neither competitor showed concern because the Roman stars were rising. The Romans were buying all the spices and distillates around, and there were plenty of coins to pay for them.

During subsequent centuries spiritus vin, in the form of distilled brandies and wines fortified with brandy, emerged as a popular trade item among the people of Italy and coastal France. The Romans perfected large-scale production of wine and its distillates.

CONVEX HELMET STILL

Rome also promoted a demand for cosmetics. As the fun-and-games, body-beautiful philosophies were exalted by the arts, young Romans gradually abandoned the ideal of leading productive lives and idleness sucked the greatness out of Rome. A nation does not die overnight; like a tree that shoots forth a frenzy of fruit when its natural demise is imminent, Roman conquerers twirled like dervishes, scattering their seeds of power to the wind. Thus, knowledge of distillation, the home production of wine, fragrant waters and medicines, as well as concepts about fermenting food, became accepted concerns among the civilized peoples of Europe.

It was only a matter of time before undisciplined tribes invaded Roman-controlled lands and the cumbersome, tax-supported Roman bureaucracy toppled. Persia, seeing the fall, struggled against Rome. Their fighting continued intermittently until both the Roman and the Persian empires lay exhausted.

Arabian Interest in Science

Inspired by the teachings of the Prophet Muhammad, the Arabians picked up the pieces and pushed their Islamic rule northward into Persia, then west to encompass Spain, south into Africa and east into China. From about 800 to 1200 A.D., the Arabians controlled vast quantities of gold plus the spice trade; the Arabic language, script, system of numerals and legal code dominated much of the world.

Remembering the Greeks who had ruled their homelands, the Arabians patterned their interest in the natural sciences, mathematics, astronomy and medicine after Greek scholarship. They fostered the use and knowledge of spices and perfected the art of both water and steam distillation.

With the increased popularity of distilled aromatic waters for medication and fragrance, difficulties in distilling some plant essences arose because of the degree of heat necessary for extraction. Plant materials lying on the bottom of the still burned easily, and with prolonged boiling, strong odors or tastes were produced. The solution was found in the steam still, which, by boiling water in a separate boiler, allowed the volatile plant parts to vaporize and condense into aromatic waters.

The Chinese and Indians were crazy about aromatic waters and would barter spices for fragrant distillates. Rose water, a twin to the earlier Persian distillate, was the most popular item. It was recommended by an emperor's physician as an internal medicament; rich Arabians used it as a flavoring in cooling drinks; in Constantinople it was considered an eye wash; Oriental chefs used it in their delicately flavored Chinese lobster sauce; and in Baghdad baths were said to have three spigots; hot, cold and rose water.

By 900 A.D. Arabians firmly established the belief that distillation was somehow a magical process of purification. The spirit of the

AN EARLY STEAM STILL

material distilled was assigned medicinal or other significant attributes. Scientists described the process of separating plant oils by upward distillation. Though they probably did not isolate chemical components of volatile plant oils, they understood that certain substances vaporized at specific times and could be collected. Separation of a liquid's constituents by upward distillation is the matrix of the fractional distillation method used by today's huge petroleum refineries. Arabian physicians of this era also put the serpentina, or worm condensing coil, into use.

I have used the serpentina in experiments with elder blossom wine; though unsoured, the wine tasted like footwash, and I was trying to remove its ugly flavor. Juices flowed through the tube all right, but in bottling my collector spilled and I lost them down the sink. I have often wondered what the little creatures in our cesspool think about our effluent.

Another Arabian contribution was the practice of "downward distilling," which corresponds with our modern destructive distillation used in the production of tar oils from plant materials. Organic material was placed in an alembic with a beak on the bottom. Heat was applied (sun or hot coals), and volatile oil "sweated" and ran with other plant liquids out the beak into a collector underneath.

Early Arabian distillers were no doubt excellent innovators and chemists. Unfortunately for science they became involved with mysticism. The wealth and vitality of the early Muhammadan era became

SERPENTINA CONDENSING COIL

permeated with a faith-in-the-miraculous culture. Alchemy, the theory of transmutation, the belief in the existence of a chemical substance that could change base metals, such as iron, into gold, grew out of the practice of distilling the spirit out of materials.

Although early Arabian scientists excelled in skills associated with distillation, after the year 1000 they suffered inbreeding of thought. Enclosed in their own circles of belief, their scientific vitality lost its power to change and grow, and their intellectual leadership faltered.

During subsequent centuries, Baghdad was overrun by the Mongolian barbarians who invaded from the East, Christian Crusaders from the West pushed back the crescent and the star, Turks crisscrossed holy places, and Arabian science retreated to the monastery.

The Medieval World

It is probable that most Crusaders who marched through the Middle East by the 1300s had become acquainted with Oriental spices, citrus, and Muhammadan learning. Even though stills had been introduced by Roman conquerors, and Charlemagne had subsequently encouraged the growing of herbs as a substitute for spice imports, the returning Crusaders probably reintroduced, and over succeeding years their offspring practiced, the distilling of medicine, fragrances, flavoring and wine.

Throughout the post-Crusade feudal times, roving bands of robbers forced vassal, lord and serf to stay close to European manor houses or church enclosures. Of the three classes, clergy, warriors and toilers, the monastics carried learning forward. With ascetic discipline, monks copied old libraries, maintained hospitals and schools, carried on plant and spirits distillation, and improved agricultural yields, while lords and vassals fought the enemy and serfs dwelt upon manorial land. Although from time to time the friars and scientists got off on the wrong track (such as exploring signatures, the plant's shape and color as an indicator of the plant's medical virtues), scholars of the Middle Ages laid the basis for much of today's science. They cataloged plants; Gerard, Culpeper, Parkingson and others wrote herbal manuscripts, which included accurate plant descriptions as well as lore, and observed medical properties. Medieval scholars attempted to isolate plant essences by water distillation, by steam and by soaking plants in fats. They concentrated their distillates by redistillation and employed evaporation to collect bottom oils. Like the Arabian scientists before them, they understood that a plant essence could be located in different parts of different plants. They also knew that different substances vaporized at different temperatures; but the isolation of the numerous constituents in one plant essence was not commonly achieved until the late 1500s.

The clue lay in alcohol. Calamus root-flavored alcohol was often prescribed for victims of plague infection, which was repeatedly introduced into Europe by vermin from spice ships' rats. In the strongly alcoholic calamus preparation, chemists discovered a film on the liquid. With that hint they began to experiment with alcohol as a solvent to dissolve plant oils. Using gentle heat, they distilled the alcohol in which plant parts had been soaked. They discovered that many volatile plant oils could be separated from their carrier. Whether accident or inspiration, the infusion of aromatic plants in alcohol and the distillation of that alcohol to separate the volatile plant oils was a chemistry breakthrough. Flavorings, fragrances, cosmetics, medicines and flavored brandies were henceforth created by macerating the plant in alcohol and distilling the solution.

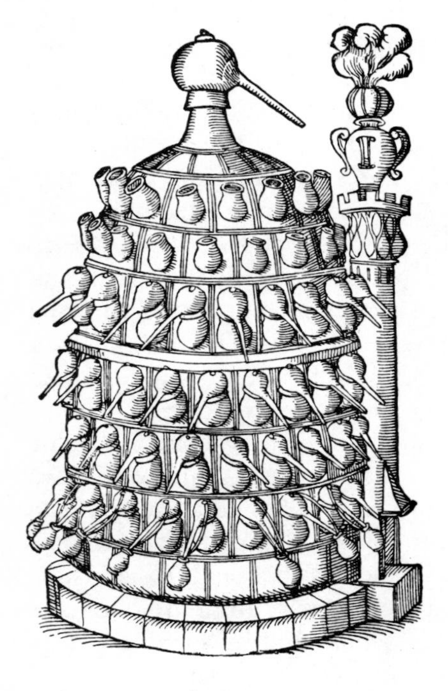

BEEHIVE STILL

A furnace is in the middle, and the retorts are placed between tiles. 1565.

With repeated waves of Crusaders flowing into and out of the Mediterranean, European maritime towns sprang up. Returning ships brought back not only infestations of plague lice and bundles of spices but also reintroduced the concept of commercial interchange in city centers where raw materials could be refined and sold at a profit. Volatile spice oils, pepper and perfumes were the most wanted articles. Self-sufficient feudal manors gave way to towns and the production of surplus goods. New systems of agriculture evolved, oat production was improved, horse usage increased, which in turn effected a better means of transporting goods, and peasants were no longer restricted to marketing their wares "at a day's walk." This led to larger market centers and the feeding of more nonfarm people.

The fear of traveling diminished, the compass was rediscovered, and guilds and money came to be accepted institutions, along with the more generalized use of the printing press. Out of these awakenings, the blind faith in authority lessened.

With the opening up of travel, people were no longer satisfied with European broad beans, bread, cabbages, onions, salt meat, dried fish, gruel and flat beer. They hungered after flavors that would reduce the saltiness of preserved meat, perk up produce, lighten drab dried foods and tease memories of tales they had heard about Far Eastern spices.

The plow, "killer of continents," increased food production, which increased population, which demanded more and spicier foods. More marketing centers evolved, towns grew everywhere, and mud, manure, litter and refuse gave off impossible odors.

In addition to olfactory offense, bad odors were thought to spread sickness such as dysentery. Town ordinances ruled that cows had to be shackled, chickens penned and pigs stied (or in some areas allowed to scavenge only at night). Some medieval cities had pig catchers, who delivered wayward swine to the slaughterhouse where their meat was processed and given to orphanages.

It is easy to understand why cleanliness and fragrance became premium values. In addition to strewing herbs for people to walk on and thus express fragrant oils, incense was burned and aromatized fats were rubbed on everything from bodies to leather saddles. The home distillation of perfumed waters was commonplace. Herbs such as rosemary, mints and sage were much sought after. Fragrant and flavorful Far Eastern spices, especially pepper, were hot items in crowded European cities.

In response, the Mongolians, who had invaded Armenia and Byzantium trade route lands, took over the caravans to China, upped the tariff and down-graded spice products. Adulteration of pepper and spices became the rule. In the 1400s, however, the Mongols were replaced by the Ottoman Turks, who cracked down on the crooks by tying short-weight artists to the tail of an ox or compelling an adulteration culprit to eat juniper berries in the amount of his falsification.

Health became a concern of medieval town authorities during the fifteenth and sixteenth centuries. Meats were inspected for lesions, which were thought to cause leprosy. Over-ripe chickens, believed to cause shortness of breath, were not only confiscated but offending merchants were hoisted upside down by their feet tied to a rope around a chimney and let loose to fly to the ground. Street refuse, the breeding ground of typhoid and rats, was cleaned by squads of prisoners, who swept the debris into rivers from which drinking water was drawn.

During this period, the study of sickness expanded into an area called pharmaceutical chemistry, which explored the properties of vegetable and animal substances as well as active chemical drugs such as sulfur, mercury and arsenic. With more emphasis on medical research, stills were venerated and laboratories spread from remote monasteries to city universities.

With the advent of printed books, wider intellectual interchange became possible and techniques of distillation were clearly described; the mumbo jumbo that had hung over chemistry disappeared. Even the great wars, which produced the formation of European states with changed concepts of land ownership and human equality, could not thwart the advancement of knowledge.

At the same time that medieval stills were being employed as scientific tools, home distillation of aromatic waters, remedies, toiletries and spirits gained momentum, fermenting of foods for winter became a common household chore, and the stillroom came into existence.

Following the Arabian influence, sixteenth century Europe moved toward specialization: Mathematics, botany and astronomy became distinct disciplines. However, apothecary chemists and home distillers, intrigued by the desiccative power of their stills, began to rekindle interest in the discovery of a miraculous cure-all. They searched once more for an elixir of health and a substance that would turn stone into gold. A revival of interest in ancient Arabian astrology spread throughout Europe. Concurrent with scientific specialization on the one hand and occult interest on the other, Egyptian and Mediterranean merchants unexpectedly put an armlock on the spice trade. Turkish wars had detoured spices from their usual trading centers, and aromatics became exorbitantly expensive. European physicians, dependent upon Oriental distillates for many of their curables, began looking for other medicinals. Almost spontaneously their pharmacists from Edinburgh to Paris turned to local herbs. The popularity of kitchen garden herbs skyrocketed; herb distillates replaced many exotic extracts, and chemical medicine, chemotherapy, gained a foothold in pharmaceutical laboratories. Alchemists could not or would not change; they continued their study of plants as the source of an elixir of health. Astrology permeated their orientations, and medicine abandoned them.

Having failed to gain pharmaceutical legitimacy, alchemy renewed its interest in metals. Research in inorganic earth materials claimed the European apothecary apparatus, which only a few years before had distilled organic materials. Gold was sought in every conceivable substance, and as the rebuffed alchemists scratched busily, their magical leanings brought theologians down on them.

AN ALCHEMIST'S STILL

All this time European states were vying for land; wars increased, revenues decreased, and plagued by money troubles, many noblemen supported alchemists in their quest for gold. The result was clerical condemnation. Throughout Europe sides polarized, and threats and counterthreats regularly erupted into conflicts. Accommodation was achieved in the end not by war but by an increased availability of spices, plus an influx of gold.

The Portuguese had sailed around the horn of Africa, opening a sea route to India and destroying the Mediterranean spice bottleneck. A triumphant da Gama had returned to Lisbon to whisper that Indian princes were eager to trade their cinnamon, ginger, pepper and cloves for European riches, which included grain distillates from Scotland. Shortly thereafter the Spanish sailed west to the Americas, where they discovered hot Capsicum peppers, allspice, vanilla and maize. In addition to the spices and corn, the conquistadores found gold; the fabulous Aztec gold was to be had for the taking. The Spanish also brought home knowledge of a strange fermentation process that the Aztec priests had effected by pressing juice from agave (century plant), honeycomb and cactus pears and incanting prayers while the juice "came to life," creating a slightly alcoholic drink.

The fruits of exploration and colonization brought comforts to Europe; they also skimmed off much of Europe's discontent, underemployment and overpopulation. Gold had a way of settling tempers, spices alleviated dullness, and all Europe thrived.

American Rumrunning

Early colonists of Virginia fermented their bread grains into spirits because they did not like distillates made from Indian maize. In later years wild grape ferments were distilled into brandy, as were the European grapes that Thomas Jefferson introduced. Rye lent its grain to the mash barrel; barley was grown, malted, fermented and processed into "uisge" (abbreviated Gaelic for whiskey). Apples contributed their essences to distilled jacks, and Bourbon County, Kentucky, perfected a distillate of maize. But of all the competing spirits, early America's most important beverage was rum.

The islands of the West Indies—Cuba, Jamaica, Haiti and Puerto Rico, plus numerous lesser islands—contain more than 100,000 square miles. Native Arawak and Carib peoples were nearly exterminated by the Spanish who had scoured their lands for gold. Some Spaniards were farmers, however, and they introduced sugarcane, tobacco and cotton. After trying to enslave the remaining natives, they gave up and brought African slaves to the West Indies. Their sugar plantations prospered because Europe's sweet teeth were willing to pay for molasses and its distillate, rum. In time rum became the avant-garde beverage of Europe.

When Dutch, French and English ships ventured to the Spanish islands in search of new markets and trade goods, the planters said that they could use more workers, but that they were not allowed to trade with non-Spaniards. That frustrated the ship captains, so they armed their vessels and began plundering the rich West Indian plantations. These sea rovers lived in hope that one day they would privateer a Spanish galleon laden with Aztec gold.

Sporadic fighting ensued until England and Spain drew up armadas. The battles ended with the destruction of the Spanish fleet and in the 1620s the English, the Dutch and the French began occupying the lesser West Indian islands. They used their toeholds as bases for their triangle trade. Molasses was transported to Europe, where it was distilled. The rum distillate was carried to Africa, where it was traded for Negro slaves. The merchants sailed back to the West Indies to sell their Negroes. Before long a French colony of *boucaniers*, literally "meat driers" who hunted wild cattle and swine, settled on the western tip of Haiti to act as suppliers for the marauders and pirates who prowled the defenseless area. With the Spanish subdued, there was no clear-cut master.

Into this sugar-rich melee New England skippers sailed, and for more than a century Yankee shipowners carried rum from Plymouth to Africa, where they traded for slaves before slipping back across the Atlantic to the Caribbean. Because of the terrible hardships and vicious

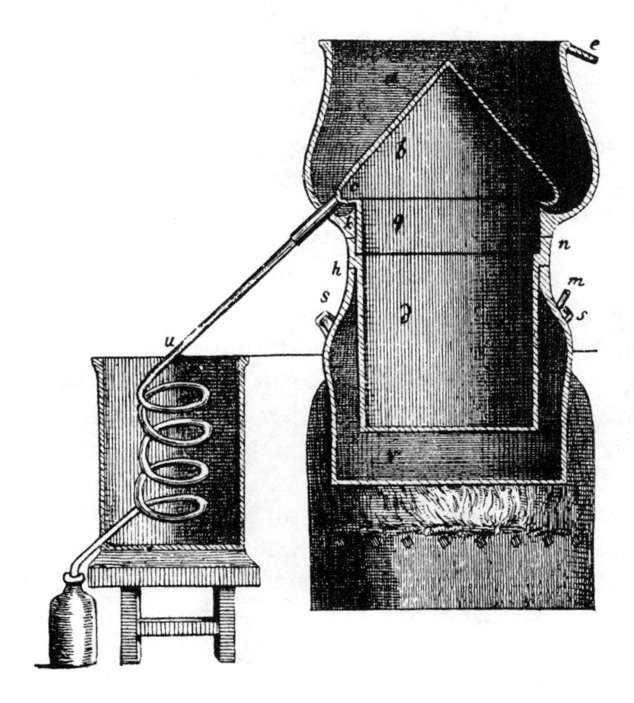

A WATERBATH STILL USED IN MAKING RUM

competition among the pirates, the death rate of black slaves was incredibly high. West Indian planters sought more and more Africans, and Yankee merchants obligingly exchanged their cargo for molasses before high-tailing it to the New England mainland where the sugar residues were sold to distillers. Distilling fermented molasses into rum—using the new technique of suspending the still cooker in a boiling water bath in order to deter scorching—helped to support the financial backbone of early New England industry.

Not all rumrunners were men. There is a story about a lady pirate, Princess Devil Star, her mulatto momma (a voodoo priestess) called her. Thin as a sapling, wide-eyed and winsome, the supple Princess, who moved and smiled like a cat, put down a mutiny aboard a Jamaica-bound vessel by bringing forth her trusty copper still and perking out rum. The thirsty crew knocked themselves senseless on the spirits. The lady pirate freed the ship's captain, dumped the snockered sailors in the shallows of a remote island and picked up her momma, who encanted a rite and married the Princess and her handsome skipper. The trio sailed off to Jamaica, then a prosperous British slave headquarters, where the new groom owned vast sugarcane lands. There, the story goes, through her skill with a still, Devil Star became queen of slave traders. It has been said that she invented rum.

After her husband boiled the juice of the cane stalks into straw-colored sugary lumps, the Princess collected residue from the boiling tables. Fermenting it with barley water and yeast, she distilled her product, running it three times until it sparkled like a diamond. She later had her liquid fire stowed in charred barrels aboard her husband's ship. When he sailed away to England with his sugar lumps and kegs of rum, Devil Star, looking every bit a princess in her jewels, brocade and fur, sailed with him to Southampton.

Leaving the rock-hard sugar bales in warehouses, the captain and his Princess sailed for the slave coast of Africa, where Devil Star tempted traders with her spirits and glitter and picked the slaves she wanted.

They say that the moon always smiled on the Princess's voyages, that she had a kind heart and would wander among her chattel singing lowly with them as she doled out rum. She often said that her slaves loved her and would follow her to hell, adding with knowing eyes, "If a woman has the tools, she can do everything."

At the last, so the story goes, she did do everything, including levitating herself skyward on the tail of a whipping blue flame. A barrel of her crystal rum tipped into a fire, and witnesses avowed that her body was lifted up, consumed instantly, leaving only the fragrance of caramelized cane. In the place where her still had stood, they found a star of melted copper fastened atop the upturned coppery horns of the devil.

In New England, where tumbling rivers met the sea, the agony of falling timber foretold the coming of a distillery. Soon after the clearing of the land, warm sweet fumes pervaded the air as the dark Jamaican syrup was cut with water. The vats were warmed, yeast was added, and fermentation writhed into a hearty sweet beer. When at last the brew lay quiet, the liquid was transferred to boilers, often steam or water bath boilers with fireboxes underneath. The still heads were clamped, the beer was gently heated, and long beaks of still heads, or tubes, fed vapors into water-cooled coils. Alcoholic vapors were condensed and their clear spirits rectified into rum.

During the winter it was written that steam rose a mile out to sea from distillery run-offs. Stories were told of fishermen catching inebriated fish, of stills blowing up like geysers and townsmen rushing to catch the spirited beer in wash tubs. Most persistent were tales of fermentation vats springing leaks. One provident Yankee awoke one night to find his town silent as a tomb; going outside he stepped into a stream of fermenting molasses. Slipping and sliding, he made it to his well where he knocked down the board enclosure and quickly directed the brewing sweet liquid into the deep hole. For the next day his still ran hot as, bucket at a time, he distilled until his well went dry.

Hearing that story, I wondered what the Yank had done with his still residues, the bottom three-fourths of watery ferments. My uncle's neighbor in the 1920s was said to have fed his residues to his hogs, but he had to quit when the pigs came down with the hiccups and their noise created a panic in the stable next door. The horses ran down his fence and all his pigs tangoed through town.

The Yankee rum trade nearly collapsed in the 1730s because of British taxes and the law requiring all molasses imports to be purchased from the Brits. Need being the mother of invention, the rumrunners designed sleek, swift sailing vessels that could out-run British policing ships. As a result Yankee shipmasters were able to take their place among the great traders of the times and become a self-reliant people who stood up to the English and their unfair tax policies.

As Americans began to filter west, spirits found their way into all walks of life. Brandies graced the tables of coastal mansions, corn whiskey warmed the hearts of lonely men, applejack gave courage to backwoods homesteaders, cognac lubricated French whistles, and rum destroyed Indian nations. From the prosperous Eastern farm to the poor plainsman's home, a corner was set aside for the still, which merrily perked out spirits as well as fragrances, flavoring and remedies of every denomination. Blossoms, berries, fruits, corn, squash, saps: All gave their essences to distillates. Americans developed their own secret recipes, giving them such names as Moose Juice, the Deacon's Spleen, Lock Jaw and St. Ann's Rot Gut.

AMERICAN HOUSEHOLD STILL

Distillation Today

Although the distillation of spirits in the contemporary home is now illegal, it was not always so. It seems that our forefathers had a great thirst for spirits. With the advent of indoor plumbing, whole populations interpreted the bathtub to be a dandy fermenting vat. Commercial distillers took a dim view of this development and in 1919 advised lawmakers that home distillation of gin was unhealthy. In the interest of humanity the Treasury Department was directed by Congress to defend Americans' plumbing. The making of distilled spirits without reporting it, complying with regulations and paying the tax was considered illicit. Unlike sneaking around with your best friend's wife, hiding the heat, fumes and apparatus of distillation was difficult to achieve. Under the law, the distillation of spirits by an ordinary citizen became illegal.

The Distilled Spirits Tax Revision Act of 1979 (Public Law 96-39) details and updates the law. The offense lies not in making hooch but in the failure to abide by regulations and to pay the tax. "It ain't no good making no whiskey, no more," an old-timer complained to my husband Lewis and me when we paused by the side of the road to help him start his geriatric truck. We live in the wilds of Virginia where there are numerous nostalgic tales of the good old days.

The art of distillation has come a long way from distilled waters, extractions and alcohols, but perhaps the most brilliant stars of the distillate galaxy are yet to shine: fuel alcohols. There are two major kinds: ethanol, which is distilled from fermented grain, and methanol, distilled from wood, coal and natural gas.

As any wino can tell you, drinking green liquor (unaltered, raw ethyl alcohol) can give you a head that feels like Mt. St. Helens on a bad day. Drinking methanol can kill you.

Many farmers, after securing a permit, produce ethanol by growing and fermenting feedstocks and distilling the fermented liquid for fuel use in their farm equipment. Because grain ferments rarely contain more than 12% alcohol (the little yeasts destroy themselves in their own juice), ethanol must be distilled several times to produce a combustible liquid. Even though big business is gearing up to produce methanol from natural gas and coal (which may be processed to produce up to 80% methanol), small-scale production of fuel ethanol is being encouraged by the United States National Alcohol Fuels Commission because most fuel alcohol is made from a renewable resource.

Because I am constantly looking for new sources of hope, new foods to grow or new ferments to brew, I decided one year when September oaks spread acorns along the forest floor to harvest a few bucketsful. The sun lay yellow across the Blue Ridge, and hot from hiking I hung my shirt on a limb and continued picking up acorns and putting them in the buckets. As I meandered deeper into the woods, eyes to the ground, stooping, picking, I did not notice the sun settle behind a haze, and when my two buckets were full I turned to retrieve my shirt. My husband, Lewis, and I have a standing date to meet at the pool for beer and jabber at 5:30 each day, and my inner clock told me it was time to go home. Unfortunately I did not know which way to walk and darkness found me in boots, pants and brassiere, carrying my two buckets of acorns, still tramping in the woods. Treetops, shrouded with a veil of fog, stood topless as if conspiring to make the forest bleak and gaunt. I realized that I was lost, but as I marched I prayed that the eye of heaven would somehow guide me. Instead of direction, a feeling of cold scales passed over my bare shoulders and a cold knocking at my heart. Tears welled in rebellion against my stupidity. Still, bucket in each hand, I pressed on.

When I topped an incline, I suddenly made out a light cutting the overvaulting darkness and, like a plane on automatic pilot, I bounded toward it. Within a few minutes I stood above a clearing looking down on a garden patio in which a birthday party was in full swing. Ice tinkled, groups buzzed in conversation, and celebrants, obviously filled with the pleasures of accommodation, crowded around a white-haired lady standing next to an enormous cake. Before I could gather my wits and slither out of sight, three dogs spied me. Barking shrilly, boiling up the hill, the hounds immobilized my flight as all eyes below betrayed their shock at seeing a woman in boots, pants and brassiere holding two buckets. People are made of colors. That gathering appeared ghastly blue. However, a gentleman came forward and called off the dogs. Though a bit uncertain about my acorns, he questioned me and offered to drive me home, a distance of several miles.

Blushing, ill at ease, I crossed the drive to his truck. I wanted to escape, but the lady with the silvery hair approached. Her eyes glowed and a good natured laugh was on her lips as she said, "Blushing is a sign of an honest heart," and handed me a piece of cake.

Looking me squarely in the eye, she inquired what I intended to do with my acorns, and when I stammered that I would try to make flour, or ferment them, or something, the lady instantly responded that her father's father had brewed an acorn alcohol. I could not believe it, but she was emphatic, saying that as a child she had picked up white oak acorns and her grandfather had malted, roasted and fermented them into beer before distilling the brew. "We had a licensed still on our farm," she added with a touch of pride.

Fermented acorns! That lovely lady did not heed my raggedy bra; we could have talked an hour, but Lewis would be worried, I said. Clasping my hand she confided, "My guests will be curious; so many people live off the acts of others, they do not understand that to do, whether it be collecting acorns in the night or fermenting acorns, doing is happiness and hope. Go brew your acorns."

Years later I ran across the following analysis concerning the composition of grains and nuts.

Name	**Carbohydrates** (sugar, starch, etc. that could be converted into ethyl alcohol for fuel)
acorn	50% to 65%
corn	65% to 67%
wheat	65% to 70%

Acorns could very well become a fuel crop of the future; it would be a farm crop that would not gully and gut our countryside. Scarred land is healed with growing trees.

Working with living entities, such as plants and fermenting microbes, engenders a respect for past roots, plus a faith in the earth and God. As earth's intrinsic energy can be neither created nor destroyed, only changed in form, distillation may be viewed as a separating process in the course of change. The still is a reprocessing tool employed to extract the essence of the earth's goodness. Knowledge of stills and the princely art of distillation one day may be a life-sustaining skill in our ever-changing world.

3

CHAPTER

Fermented Foods, Staples of the Stillroom

Walking one day I befriended a thin young woman whom the insults of life had shrunk into indifference; her response to conversation was a shrug. Reaching home, I gathered a few sandwiches and we picnicked among the rocks of the North Pacific shore. After eating lustily, she laughed a frosty little snicker and admitted that she had loved a man, had given everything, and that he had disappeared when his seiner lost power over the halibut reefs. She said she had been ashore on the radio-telephone as he raged against the world. "Everything is a sham, only food is real!" were his last words.

The girl added that two wives had come out of the woodwork to claim his "benefits," and they took all. Resolution glowed in her pale gray eyes when she asserted, "A full belly is the greatest gift."

Though I understood that her impudence was rooted in fear, that merciless machine that swallows hope, I disagreed. "Courage, a trust in God, love, are the earth's secret ferment, they are man's greatest gifts," I said loftily.

"Food, like money, ain't important unless you have none," the girl threw out.

To the present day, as I delve into the history of foods, the image of that hungry young woman rises up before me.

If there is a common denominator of all mankind, it is food. From the earliest days, when man considered any substance that could be beaten into a pulp and digested as fair game for his gullet, to the most exacting contemporary gourmand, food has dominated the altar of man. Ferments are as old as the living planet, but say the word ferment and people launch their brain cells into spiritus orbits. Actually, throughout the globe the fermenting of foodstuffs is a gargantuan occupation as compared with the fermenting of alcoholic drink. On a reptilian scale, the fermentation of food is a boa; booze making is a pencil snake. Fermentation is a chemical changing process, during which food is altered but nutrients are usually retained; it is an energy-efficient way to enhance and preserve food.

Future Foods and Microbes

After the ice retreated, warm winds encouraged wild grains to sprout and civilization took root in the Middle East and in China. Man developed patterns of producing food that have been imitated ever since. "Strip the earth's trees, scab her skin, force her to produce food for here and now!" Since 4000 B.C. people have clustered along rivers to cesspoolate the water. They have turned once-fertile lands into deserts; they devised ingenious methods of irrigation, only to leach the soil of its growing power. In ancient times, when one area became impoverished, man could move on. Today there are no promises on the other side of the mountain. After thousands of years of farming there are few untilled wrinkles left on the old earth's skin; and the need for food is exploding.

Chemical fertilizers cannot be considered the panacea. The limited soil nutrient, phosphorous, is one big hole in the doughnut. Chemistry has not been able to manufacture phosphates, which basically come from fish. Like "The House That Jack Built,"

This is the ocean,
That swallowed the river,
That carried the waste,
That came from the food,
That grew in the field,
That absorbed the guano,
That came from the bird,
That ate the fish,
That swam in the ocean . . . We are running out of guano.

Future foods must not destroy the earth's precious skin. They must be tempting yet nutritious; they must have keeping qualities and conserve finite energy, and they must reproduce at a faster rate than people. Future foods might well rest with microbes.

The art of fermentation, that is, the changing of organic compounds into various foods and beverages by means of bacteria, yeasts and fungi, is ancient knowledge. Fermentation, the matrix of spirits, not only feeds the womb of the still, it is the mistress of earth life; ferments both give and destroy life. Over the years we have learned to harness ferments, which we employ in bread, cheese, kraut, sausage, pickles, soy sauce, yogurt, fish, pickles, coffee, tea, cocoa, tempeh and sufu; but the heyday of ferments may be yet to come. Scientists are developing single-cell microbes that will provide a direct source of human food. Zillions of these little ferments, which are grown from carbons, may be dried into nutritional tidbits. At present all that is lacking is an acceptable flavor. Yuk! my young friends who tried the super food described its taste; yuk, yuk when they tried it in milk, triple yuk on pizza. But flavor will come; an empty belly is a great motivator. Proximity to hunger breeds acceptance.

Bacteria, Yeasts and Molds

As in a three-ring circus, food fermentation develops in a trio of circles; the bacteria clowns, the yeast hot shots and the highly specialized mold professionals all perform when conditions are right.

Like dwarfs of the big tent, lactic acid bacteria should get into cheeses, yogurt, pickles and sausages before harmful bacteria. Strains of *Lactobacillus* are small and have relatively more surface with which to quickly devour their host's carbohydrates and produce an acid that stops many other microbes cold. There are dozens of lactic bacterial cousins that do not need air, so they penetrate and feast unseen on milk, vegetables and meat. Cheese, yogurt, kraut, pickles, sausage and ham are made by the invasion of microscopic lactic bacterium clowns that are introduced from soil, water and air. Some are droll, sad-faced bacteria that produce milk acid, some are tumbling slapstick characters that convert sugary constituents into strong lactic acid, others are hoary roustabouts that produce lactic and acetic acids. All eat sugars, grow, divide and give off acid until it inhibits their own growth and a more tolerant species of microbes take over. A clown can create just so much bitter laughter before his antics cease to provide support.

Yeasts that feed on sugars are the hotshot experts of bread, brew and wine arenas. Like human cannon balls, they give exhilarating highs, are often filled with their own gassy importance, and need air to survive. Like their microscopic neighbors, they are gluttons. Yeasts eat and divide until they find themselves victims of their own ethyl alcohol and carbon dioxide excrements, which soon put them out of business. If yeast action is not stopped, these daredevils of the microbe world will blow themselves up.

The highly specialized mold ring in the food fermentation circus produces outstanding visible acts. People can see molds. By their very specializations, molds, which can convert grains and beans into easily digestible tempehs, create flavorful cheeses, synthesize vitamins, and grow into mushrooms, put themselves into jeopardy. It is easy to point the finger at things you can see. Molds give off toxins; they can poison and spoil food as well as enhance it. Some molds take on the appearance of kindly white fuzz or dance on lawns in a mushroom fairy ring, when in truth they harbor deadly enzymes; others look devilish but are harmless. Their very name, mold, conjures up dual images.

Both mold foods and yeast-fermented foods need air, moisture, nutrients and warmth to flourish. But left alone they will encircle their world, decompose, and make the food they infiltrate unfit to eat.

Fermented foods may be the result of the acids produced by lactic acid bacteria, or alcohols and carbon dioxide produced by yeasts, or mold growth that breaks down carbohydrates, plus secondary acetic acid bacteria that oxidate alcohol into vinegary acid. Fermented foods may utilize a combination of these microbe processes that cause enzymes to change organic substances into different foods.

Because most lactic acid bacteria do not need oxygen, cheeses, naturally pickled vegetables and fermented meats do not readily rot; their acidity retards invasion of other bacteria that could cause spoilage. Scientists say there is very little calorie difference between the starches and sugars of the original food and the lactic acid components created by fermentation. When the circle of microbe activity is stopped, there is little nutritive loss. In fact, some fermented foods contain higher food value than the host substance.

However, like many human circles, fermenting arenas may deteriorate into sluggishness. Ferments can develop a credo of misunderstanding resulting in slimy, acid bitterness. Stopping microbe activity by the proper action is as important as introducing new blood into human circles.

Microbes, like teensy weensy people, need food, water, warmth and often air to survive. Take away any one of their life supports and fermentation is slowed or stopped. The art of fermentation lies in feeding the microbes their favorite menu, keeping them moist and toasty, and providing oxygen to those that require it. If a microbe's food is contaminated with too much alcohol or acid, it cannot survive. Drying food stops a microbe life cycle by taking away its water. Boiling or freezing kills most microbes, and vacuum packing removes the air, thus destroying air-using microbes. Learning to stop fermentation at the precise time to maximize keeping qualities, plus the product's tastiness and nutrition, is the aim of a food fermenter.

Purists say that fermentation shortcircuits the potential energy inherent in carbohydrates, proteins, fats and nucleic acids. True, but the very act of shortcircuiting results in a preserved often ready-to-eat product.

There are toxic microbes. Aflatoxin may be produced by mold ferments; botulism poisoning is caused by a bacterium that forms in vacuum-packed food; some streptococcus strains can cause illness and death. Heat, pasteurization and high acid solutions can combat toxin producers. Thus with proper fermentation, sealing and storage, fermented foods are not hazardous to health.

A Brief History of Fermentation

Fermentation of food was practiced by Egyptians living 3,000 years before Christ. When sheep and goats were domesticated in dry lands and water buffalo were tamed in wet areas, their milk was fermented into cheese. The Egyptians also understood that fruits, left untreated, turned into edible vinegary products and that salted olives turned into pickled olives. Egyptian women discovered that the gas generated by itinerant yeasts made their gluten-grain breads rise and that the leavened breads kept better than traditional fried breads.

Though each took a different form, civilizations of the Indus and Yellow River valleys apparently fermented foods concurrently with the people of the Nile. Some East Indian peoples of these ancient times lived in elaborate riverside cities complete with granaries; they had markets for legumes and green vegetables, some of which they sold fermented in animal bladders; they had dairy shops complete with goats for fresh milk and leather bags filled with coagulated products. These people sold jungle fowls, ancestors of our chickens, and some merchants dealt in cattle. Flat-backed Indus bovines, which were bred as super animals for draft and milk, provided them with fermented cheese foods. A fertile cow could feed a lot more people than its carcass, wise men said; thus religious and cultural dictates discouraged beef eating. Ancient bacteria contributed to a variety of curds with good keeping qualities, which added protein to diets.

In contrast to ancient India's reliance on lactic acid bacteria to ferment milk products, the Chinese who crowded into the Yellow River Valley several thousand years before Christ regularly fermented whole tiny fish into salty fish sauces. By use of enzymes in fish digestive systems, the pungent sauces flavored and added food nutrients to millet and beans, which were preponderant in Oriental diets. Rice, which originated in the wooded marshes of what is now south China and Thailand, was later added to the millets and beans, and all three staples were complemented with the powerful fish ferments. Rice itself was used in fermenting sauce. During the Chou dynasty, around 1000 B.C., the fermentation of brined soybeans together with other grains, molds, yeasts and bacteria was an ordinary household chore. These ferments, which produced soy sauce, were used in seasoning and in enhancing flavor. (Monosodium glutamate, MSG, is a by-product of similar soy sauce fermentation.)

Each summer Chinese peasants moved from their winter huts to the dikes that cut the vast fields in which they labored. On the plains, fuel was scarce and cooking was restricted to the quick-frying of thinly sliced vegetables. These were usually supplemented with precooked rice or millet pancakes, which were brought every week or so from their winter homes. To lessen the monotony of vegetables, boiled rice or cold pancakes, protein-rich fish and soy sauces, which had been fermented the previous winter, were carried by each family to their summer camp. Even in the tiniest amounts, those concentrated sauces added pleasure to bulky diets. During the following centuries fermented soy sauces and pastes added variety to pork dishes made from the miniature pigs that were part of rural households.

During these same ancient times, Southeast Asian peoples concocted tempehs, consisting of precooked grains and beans that had been inoculated with edible molds, which were thought to come from household straw. After short fermentation periods the tempeh cakes were quick-fried in oil and eaten as a staple.

Fermented cereal or bean foods were attractive to ancient peoples because fermentation very often improved the flavor; the nutritive value of mold-fermented foods was increased; and the digestibility of treated grains was easier than that of plain cereals. Fermenting beans and grains with molds to form tempeh reduced the cooking time required, and if the product was brined, the secondary fermentation increased the keeping quality of the food. Fermented tempehs were instant meals.

Lactic acid fermentation of vegetable substances, though known throughout antiquity, was popularized and expanded during classical times. The brining and fermenting of olives, cabbages and cucumbers were elevated from home arts to commercial production by the Greeks. Vegetables were subjected to a succession of brines that withdrew natural sugars, and the sugary solution became food for airborne microbes. The fermenting bacteria and some yeasts gave off lactic acid together with a bit of alcohol, which attracted other microbes and produced an acid bite to the krauts and pickles. Cabbages were the kings of kraut, but mustard and other greens were also fermented.

During the summer of 1949 I substituted in a primitive village school across the Bering Straits from Siberia. There were no school classes at the time, but government property, the radio-telephone and plane mail had to be cared for. In one of the outlying, half-underground houses lived a deaf man named Petr. Endowed with massive shoulders and short legs, the Eskimo rowed his kayak with furious energy, or untiringly wielded his ax on driftwood until the muscles of his arms nearly split his jacket. Hardly ever friendly with the villagers, the man wore a forbidding manner as he went his silent way; even the dogs were quiet as he passed. I once saw him grab a cat and swing it around his head before letting it fly. Though I had no personal reason, I was afraid of him.

The village people planted no gardens, but women and children gathered pot herbs such as willow herb, coltsfoot, roseroot and sorrel, which they boiled as a vegetable or fermented into kraut.

One evening I was a distance from the village gathering edibles when, to my shock, I looked up to see Petr standing nearby holding out to me a sack of greens. A smile of a simpleton lit his flat, broad face. I did not know whether to yell or to run. After a momentary silence, I greeted him, said thank you but no thank you, and turned toward home, seen only as a haze of smoke in the distance. No person was in sight.

Petr followed me and if I looked his way, he smiled. When he grasped my elbow I nearly collapsed. Making inarticulate sounds, he pointed to a small bird faking a broken wing in an effort to lead us away from her nest, and with his huge hand he parted the bushes to show me the nest with speckled eggs.

By the time we reached the village shore, my fear had left me. Though I did not take any of Petr's greens, I watched him chop and bruise them before stuffing them with salt into a seal bladder and tuck it into the stilted cache behind his house.

Later I learned from villagers that during the long grim winter Petr would eat his sauerkraut "like a woman." Men of that village did not eat vegetable food. Petr, mimicking that which he had seen his mother gather and prepare before she died, probably never had been "told," nor never knew that krauted vegetables were woman's food. The man's act of gathering and making sauerkraut was greeted with contempt by village males.

I felt sorry for Petr and yearned to somehow explain to him that men of his village did not gather nor eat sauerkraut. I did not know how to tell him.

Fermentation using airborne bacteria and yeasts was an accepted household chore during the early Middle Ages, when broken social systems and invasions caused all Europe to retreat into self-sufficiency. With Rome's protection and trade organizations kaput, food again became seasonal and the need to preserve it became all-important. The rural kitchen became the heart of the home, and the big pot hung over a fireplace became its soul. Because there was little iron for building ovens, nor wood to heat them, puddings made of grains or legumes were tied in a cloth, steamed over the caldron and eaten day after day—Peas porridge hot, peas porridge cold. . .

There were few spices; only fermented cabbage, cucumbers, turnips and beets broke the terrible monotony of drab food. During those dank ages of isolation, inhabitants of cottages, leery of robbers, learned to covertly dry grain and beans, to ferment a few pickles, and to salt down surplus fish and small game. Domestic animals were scarce because thieves lurked in every ravine ready to snatch, cut and run. The rural cottagers hid their food in a secret corner.

As a defense against outlaws, castles with protective moats were built by noblemen and monastics guarded themselves behind walls. Seeing strength in numbers, peasants clustered close to the strongholds, and in return for protection they willingly paid tribute in puddings, pickles or wine. In time the voluntary gratuities became levied taxes and the enemy included other princes trying to muscle in on the action. Families were obliged to donate a sturdy son to fight for the manor lord or a maid to weave his clothes. Wars endlessly rocked the balance of power between princedoms, but gradually Europe stabilized. Gardens attended rural homes, cows were pastured without rustlers, pigs and chickens graced every village, and with increased food varieties, methods of preparing and preserving foods expanded. Each geographic area had its own versions of fermented cheeses, wursts, wines, breads and pickles.

When European peoples were called upon to unite and rescue Jerusalem from profanation, all classes gathered their preserved foods and marched to the Holy Land, where they discovered spices and the Islamic still. Brought home, the still became a status symbol, a show that one of their own had marched with Crusaders. In manor houses the still was enthroned in a room with "noble foods:" choice cheeses, wines and other fermented products. Out of pride the place was called a stillroom. The rich had their own distiller; monasteries, universities, and apothecary shops had their specialized stills; and down on the farm, people imitating their "betters" maneuvered a few perked-off spirits, medicines and scents, which lined shelves together with their fermented foods.

Fermentation of Coffee, Tea and Cocoa

Although I grew up fermenting all kinds of culinary mischief, I was not conscious of the universality of fermented foods until I read about a Ukrainian Fermented Foods Festival, which must have sent participants home groaning with pleasure.

Pickled beets with florid onions
Cheesy herring fermented in their own enzymes
Breast of veal marinated in whey (roasted until pink and garnished with dill flowers)

Lavish Danish ham
Horseradish mayonnaise
Crusty triangles of bread slavered with sour cream butter
Smokey Greek olives
Black bean pancakes made with tart curd
Regional caviars
Rich cheeses
Hard sausages and pale wursts
Fermented vegetables with hot peppers
Corned beef brisket slivered with shavings of pickled ginger
Cucumbers stuffed with blue, cream and cottage cheese
Smoked salmon
Sauerkraut and pickled knuckles

Fermented foods are alive and as provocative today as they were centuries ago.

For example, our coffee, tea and cocoa are produced with the aid of fermentation. The coffee craze spread from Red Sea countries, where coffee growing was a thriving industry in the 1500s, to Brazil, Colombia and other South American countries. Bushy plants, such as *Coffea arabica*, bear jasmine-sweet flowers that develop into two-seeded red fruits. The hand-picked coffee berries are thinly layered on trays in the sun, where spontaneous bacteria and yeasts ferment their pulp. These eager microbes assist with the drying process, adding flavor and fragrance to coffee beans.

Black tea leaves, *Thea sinensis*, undergo an oxidizing fermentation after they are picked, briefly wilted and run through rollers that rupture the leaf cells and express juice on the tea leaf surfaces. The leaves are then spread on trays and stored in darkened sheds where tannin changes their color to near black and yeasts ferment the surface juices to enhance the odor and taste.

Cocoa beans grow on trunks of tropical American trees, *Theobroma cacao*. Both the Mayas and Aztecs learned to cut open the large cacao fruit pods, scoop out the pulp with its beans, and allow the pulp to ferment for about a week. Contemporary processing still involves yeasts eating the sugars in the pulp, converting them to alcohol, and bacteria moving in to form a brown acid liquid that helps to color and flavor the cocoa beans. Without fermentation, the three beverages that are so much a part of our diets would not contain their characteristic flavor and aroma.

Bacteria in Cheesemaking

Start your own stillroom in the crisper drawer of your icebox by fermenting cheese. Cheesemaking involves: pasteurized and cooled milk; a starter of sour milk; sour cream or a commercial starter, which may be strains of *Streptococcus lactis* or *Leuconostoc cremoris* bacteria; an incubation period at room temperature to sour the milk; rennet as a coagulant; clabber heated for an hour or so to separate the whey; clabber poured through a cheesecloth; rinsed solids; and drained curd. The curd may be salted and eaten as an uncured cottage cheese or it may be cured by a second bevy of lactic acid bacteria or other ripening microbes. Rennet and fermented milk cultures may be purchased from Nichols Garden Nursery, 1190 North Pacific Hwy., Albany, Oregon 97321.

To pasteurize raw milk, heat the milk to 145°F (62°C) in a double boiler and hold at this temperature for 30 minutes. Stir while heating. Cool the milk as quickly as possible.

Most cheeses are made from cottage cheese that has been salted and pressed to squeeze out the whey, salted a second time, inoculated with microbes and wrapped in cheesecloth to ferment or cure. The initial curing is usually done at room temperature followed by a maturation period in a crisper drawer, about 50°F (10°C). When the curing time is up, cheeses may be dry salted, brined, coated with paraffin or eaten.

Swiss or Gruyère are given a special shot of bacteria with their starter. That causes these poor curds to develop gas, which produces their holes. I must have eaten some of that bacteria once at a fish fry. I was slated to give a demonstration on ferments, but my tortured innards ballooned, a kindly soul gave me some epsom salts, and the only thing I demonstrated was how fast I could run into the woods. Lewis later cured me with a cold beer.

Cheeses are grand to eat at any step along the way; like shiftless people, cheeses can live without principles. If the curd is hard and conservative, use heat; it is great for pizzas. If the curd turns out to be soft and liberal, spread it. If your cheese matures to be just right, share it with one you love and toast the health of friendly microbes.

A man whom I met in Alaska made cheeses that he said he preserved with a vegetable oil rub. One does not meet many cheesemakers, and I remember him as a very tall, former Army cook who had the profile of a horse. He parted his wheaten mane straight down the middle. It was easy to see that he was head over heels in love with his hair and his cheeses; he cooed over and cultured both loves. He married a capable cook whom he sent out to work each day while he stayed home and dreamed of opening a cheese factory. Life is made of dreams, but they must be given substance by action. I would have felt better about his dreams if there had been one cow in the area.

Old horse-face made a good Romano-type hard cheese, people said. I did not taste it, but I know that his wife, our school cook, made a fine cottage cheese by filling the big iron skillet with milk, adding sour cream and putting it into the pilot-lit oven overnight. The next day she stirred in part of a rennet tablet that had been dissolved in water, returned the skillet to the warm oven and left it undisturbed for a couple of hours. Finally she heated the curd on the stovetop over a low, low fire, all the while stirring and breaking the curd with her fingers. When the curd was hot-to-the-touch she turned off the heat. She separated the curd from the whey by letting the whey fall through her fingers. When it cooled, she poured it through a cheese-cloth-lined collander to drain, washed it with cold water and drained it again. "Clabber is honest, straightforward food, ferments do all the work," our jolly cook used to laugh when we complimented her on her cheese. We loved to see her laugh, she was a person who brought joy into a room when she entered.

Her clabber cheese was sometimes inoculated with blue cheese for flavoring before being crumbled into salads. Sometimes she added mint and a touch of olive oil, or sometimes the boarding school children begged her to dredge spoonfuls in flour and fry the cheese. Finger food, it came by the yard when the students bit into the patties. How they laughed, she laughed with them, and the whole dining room lit up.

Bacteria in Pickling

If you want to leave a lasting inheritance to your children's children, "plant a capers plant," an Italian told me during lunch in a vineyard. We were talking while eating our sandwiches and a scattering of capers had fallen from his bread and salami. I was trying to earn tuition money by picking grapes, only I am horrified by spiders, and huddled under the arbors with me were hundreds of eight-leggers ready to pounce. By the time I had examined every stem and filled my crate, the sun was sinking. I made 20 cents on the box, which was not enough money for bus fare back to town. My Italian co-picker, a ponderously thick man with barrel-shaped legs, a yellow face and a sluggish walk, said I could ride in the back of his truck as far as the trolley line. From there, my 20 cents would get me home.

I eagerly climbed over the tailgate with the other pickers, noticing one girl with large dark eyes, a charming smile and dazzling black hair. We sat together and soon were jabbering like old friends. The capers philosopher driving the truck was her husband. He had not spoken to her since they had lost their only child in a freak accident the previous year. They had been at a picnic near a mountain lake with their Bible school when their three-year-old climbed into a rowboat, floated from shore and was mid-lake before they noticed him. The man could not swim, but my acquaintance said she waded and floundered to the child, where she accidently tipped the boat and her son drowned. I was shocked by her story, it seemed to be so flatly told, as if all emotion had been wrung out of it. I started to offer condolences, but at that moment my bus stop appeared and I had to pound on the cab, a signal to stop. As I climbed over the side, swinging forward to thank the driver and glancing back again to say, "I'm sorry," to his wife, the girl giggled prettily, "He thinks it was his, but it ain't." Her magnificent eyes lighted as the truck pulled away from the curb.

Capers, *Capparis spinosa*, grown in the Carolinas and south, may be changed from acrid buds to pungent pickles by simply brining them 40 minutes in 20 parts water to 1 part salt, draining and dropping them into a bottle of boiling vinegar. I have employed the same pickling method to make counterfeit capers out of green elderberries, picking and pickling them just before the berries turn dark. Real or counterfeit, pickled capers may be the fastest pickle in the East, but whenever I bite into their tart hearts I cannot help thinking of those troubled grape pickers.

A second real-fake pickle is Pickled Hot Peppers; like fake capers, they contribute a snappy nip to sandwich joy. Prick ripe red, yellow or green hot peppers, *Capsicum*, with a nut pick and allow to stand overnight in a 20 to 1 brining solution to cover. Rinse, drain, pack into jars, cover with boiling vinegar and seal. I sometimes poke a handful of dill or a few crushed garlic cloves into each jar and use the pungent vinegar as an additive to salad dressing.

Fermented pickles may be likened to shoulder pads on a man's well-tailored suit; are they really his? Wow! And a woman's educated eye follows the profile to his hips to measure proportions. So also does a shopper study a pickle jar. Lactic acid bacteria in a naturally fermented product will cause sediment. Look on the shoulders of the pickles or on the jar bottom for the validating whitish dusting. One warning: with today's technology I have discovered that the detection of fake pickles, or padding, is difficult.

Pickled foods are instant foods; once preserved they require little processing. There are various ways of pickling vegetables. So-called pickling may be achieved with the exclusive use of vinegar and/or chemicals. Vegetables that are fermented in a weak brine for a short time and sealed or treated with vinegar, sugar and spices are also called pickled. Brine-fermenting vegetable pickles, made by the gradual addition of salt, which withdraws the plant liquids that act as a food for acid-forming bacteria, is called the long method of pickling.

Successful fermenting of vegetables relies on reducing natural plant enzyme action, raising the acidity, inhibiting bad microbe invasion and slowing oxygen-chemical changes.

In pickling krauts, olives, onions, cucumbers and other vegetables, the thing to remember is that they are low-acid foods that must be made high-acid by the encouragement of bacteria—preferably lactic acid bacteria because they deter other microbes that may cause spoilage. Acid inhibits oxidation, a chemical process that makes pickles mushy. Fermentation of vegetables is a natural phenomenon that takes place with proper ripeness of the produce, temperature, pH, cleanliness and salt concentration. The good thing about harmful bacteria is that most do not thrive in acid substances with a pH of 4.6 or lower, nor in 10% salt solution; nor can they take heat.

A few years ago I met a pickler who vowed that bad pickles were caused by air, and he employed a gallon plastic bag, half-filled with water, as his air lock. Layering it within a wide-mouthed gallon jar but on top of his fermenting pickles, he pulled the sides of the bag up and over the jar rim and secured it with a rubber band. He made great pickles. I knew that fermenting gave off gas because I nearly gassed myself one time when I swiftly lifted a lid and poked my nose into a crock; so I asked him why his pickling pickles did not blow out the plastic. He explained that carbon dioxide exchanges place with the air in the jug. The regulation of gases is an important part of pickling crisp, tangy vegetables; so exclude air, my pickling neighbor emphasized.

Most old-time pickle recipes have recognized this need to exclude air if the fermented product is to be preserved. Restricting air helps to cut down on the growth of yeasts in lactic acid ferments. That is necessary because some yeasts destroy lactic acid, which in turn does away with the pickle's crunch. If you discover a white film atop your pickle fermentation, you know yeasts are at work. A little is harmless, but skim the scum, add a light sprinkling of salt, keep your container full of brine, and make your pickling crock as air-tight as possible.

The Chinese used to brine and ferment cabbage, radishes, bamboo shoots, turnips, cauliflower, mushrooms, carrots, ginger and onions. They packed the produce in a 2% brine for 24 hours. The brine was increased to about 5% for about a week, and the salt content was gradually increased to about 10% by the end of a month. At this time the fermented vegetables were sealed with wax. If sealing was not possible, a 15% salt concentration was established to inhibit spoilage and keep vegetables from freezing. Salt was leached before use by soaking the produce in fresh water, after which the vegetables were flavored with garlic, ginger or spices.

Europeans have lengthened the Chinese pickle list to include string or broad beans, cooked beets, brussels sprouts, green tomatoes, mustard greens, okra and fresh corn.

Cabbage Relish made with squash is our newest pickled friend. My husband, Lewis, hates summer squash such as zucchini. I plant it because it is so responsive—and gardeners need response. Now that I have learned to substitute coarsely ground summer squash for the cabbage, or the tomatoes, or part of each, in the old Cabbage Relish recipe, I harbor no guilt feelings when planting ever-loving squash. Sweet Lew does not know that he is eating squash and he likes this relish.

Cabbage Relish

Or its squashy counterfeit

First day: In a meat grinder chop enough cabbage (or substitute those zucchini baseball bats) to make a quart and enough green tomatoes to make a quart. Soak the chopped cabbage and tomatoes separately overnight each in a one-quart brine, four tablespoons (60 ml) salt to each quart of water.

Second day: Fasten your food grinder out-of-doors so that you will not asphyxiate yourself and chop one cup (250 ml) red or green hot peppers, sans most of the seeds, one cup (250 ml) sweet peppers sans seeds and two cups (500 ml) onions. Lewis likes this relish with all hot peppers, no sweet ones.

Drain off the brine from the cabbage (or its squash subsitute) and the tomatoes; do not rinse. Mix, add the peppers and onions together with one cup (250 ml) sugar, a quart of vinegar, five tablespoons (75 ml) mustard seed, one tablespoon (15 ml) celery seed, ½ teaspoon (3 ml) turmeric. Mix and let stand for two hours.

Simmer the mixture until it is clear and seal at once in hot, sterile jars. This makes about five pints. Equal parts relish and oil create an instant salad dressing.

Yeast in Breadmaking

It has been said that grain was the mother of civilization. Grains provided a dependable sustenance and, in return, the growing plant extracted discipline. Men could not roam willy-nilly over the neighborhood and also tend the field. Of the cereal grains—rye, barley, oats, corn, rice and wheat—the *Triticum* family of wheats is the most universally employed as food. Soft wheats are fine for cakes, but hard spring wheat is used for bread dough because it contains more gluten, which provides the elasticity that forms a soft web and captures the carbon dioxide bubbles of fermenting yeasts. This adds airiness to bread. If other cereal grains are used in breads, proportions should be about one part odd flour to four parts wheat flour because other grains contain less gluten. If too much glutenless grain is used, the bread is likely to come out like a rolling pin; it will not rise adequately.

Although some wild microbe fermenters are present in most flours, today's store-bought yeast, *Saccharomyces cerevisiae*, is more dependable. But yeasts can trick a person; I made a batch of bread that grew to enormous proportions, and the larger it grew the more liquid it became. Finally, after numerous kneadings, I dumped it into the dishpan and added flour. When it rose out of that container, I popped it into loaves and shoved them into a hot oven without further rising. The heat stopped that colony of playful microbes; they must have been oversexed or half starved. Surprisingly, the bread was fine.

Just as bad bacteria can contaminate other ferments, bad yeasts can cause ropiness, soured bread or bread that refuses to rise. It is best to test yeast with a little sugar and warm water; if it refuses to fizz, toss it and try new yeast.

In breadmaking, fermentation starts immediately; the yeasts eat the sugars, give off carbon dioxide and alcohol, which makes the dough smell good, and the dough continues to ferment until it is baked. During fermentation, temperatures should be cozy, like a warm day. Once in the oven the dough rises very quickly with the rapid production of gas (this often causes the baker's soul, the big hole you sometimes find in a loaf), but when the bread heat rises to about 149°F (65°C), the yeasts are killed, the alcohol evaporates, and proteins and carbohydrates coagulate. Crust is caused by the sugars in the dough caramelizing.

To my mind, nothing quells the beast in man or child as readily as hot, fresh bread. Egyptian Pocket Bread, though as old as the pyramids, is as new as tomorrow.

Egyptian Pocket Bread

Makes about 15 rounds.

Activate two packages of dry yeast in a little warm water with a bit of sugar; set aside. Into a large bowl, pour about a quart (one liter) lukewarm water or a water/milk mix. Add two teaspoons (10 ml) salt, four teaspoons (20 ml) sugar and four tablespoons (60 ml) oil. Stir in the activated yeast. Gradually add about two cups (500 ml) whole wheat flour and ½ cup (125 ml) soy flour or fine-ground corn meal. This last half-cup of odd flour is optional; you can increase the wheat flour by a half-cup if you do not have the soy or corn meal. Stir in about eight cups (two liters) all-purpose flour. Flour is persnickety: The moisture content, grind and kind of grain varies the result, so hold about half of the all-purpose flour and gradually add it until the dough is tight enough to empty on a well-floured surface. Knead in enough flour to make a sheen on the dough. This takes about ten minutes.

Return the dough to the bowl, oil the top, cover and set to rise in a warm place for about two hours. After that time, punch down the dough to let out the gas, grease top, cover and for a second time allow the dough to rise until double in volume. Punch down.

Taking a wad of dough about the size of your fist, place on a lightly floured surface and roll out to about the size of a saucer; thickness should be about a scant ½ inch (1 cm). Place the rounds on an ungreased cookie sheet. Make sure to stagger them a distance apart so that they do not run together.

Let your Egyptian Pockets rise for about 20 minutes, then bake for about 15 minutes or until nicely brown at 450°F (230°C). I make up another couple of trays while the first two trays are baking, and to rush things I often turn the rounds over with a spatula for the last few minutes of baking time. If the Pocket Bread does not make good pockets and tends to break rather than bend when you cut it, use less oil in future batches. The rounds freeze nicely, so I usually make up all of the dough.

Egyptian Pocket Bread is worth the effort when you see the pleasure of young people bellying-up to a table filled with a heap of Pockets and stuffings. Either cut these flying saucers in half like hamburger rolls, or cut partway and, if the pocket is not well defined, hollow out the insides. Fill with sloppy hamburger/chili bean mix topped with fresh onion; a crisp salad with perky dressing; fried oysters; or as Lewis and I often enjoy for supper, dribble olive oil on each half of the pocket and load in lettuce, onion, provolone, pepperoni and mashed-up hot peppers. The olive oil and hot peppers makes Egyptian Pocket Bread sing Italian.

Molds in Tempeh and Other Foods

Information regarding the purchase of cultures, tempeh starters and calcium sulfate may be secured from GEM Cultures, 30301 Sherwood Road, Fort Bragg, California 95437. You might also check with a local natural food store for supplies.

Molds employed in food fermentation suffer a chronic identity crisis. Like a harlot screaming that she is a lady as she is being boosted into the paddy wagon, molds overtly support their reputation as food spoilers. At the same time "good" molds create Blue Cheese, citric acid, soy sauce, plus some terrific Asian foods made from grains and beans.

Miso, or bean paste, is made of soybeans, rice and salt, plus mold and other microbes. It may be eaten as a staple or as an additive to flavor and give nutrients to other dishes. Miso is somewhat like soy sauce in taste. I would love to try to make it, but unfortunately the flow sheets giving directions look like a corporate consolidated production chart.

Tempehs, on the other hand, are fun to create. Tempeh (pronounced tem'pay) is the name given to grain or bean cakes fermented with mold. Primitive Indonesian tempeh making involved soaking soybeans overnight in water, rubbing them between the hands to remove hulls, boiling until soft before draining the beans, drying them, wrapping a scoop of the cooked beans together with a bit of self-saved tempeh in a banana leaf, and incubating the material in a warm place for two to four days. When the beans were bound into a firm cake with a white mycelia of mold—which looks like the mold on Camembert cheese—plus a few dark spore patches, the tempeh was either sliced, salted and briefly fried, or it was chunked into soup like meat.

My grand schnozzle is tuned to odors of ferment, and to me, tempeh has an earthy scent like mushrooms. It is easy to make and cook; it is highly nutritious, it digests with ease, and all kinds of partially cooked grains and beans may be used in making it. This staple may be briefly parboiled and frozen, it may be dried or brined for long keeping, or kept in the icebox for a few days. Though it absorbs flavoring from other foods, plain tempeh, like a young girl, has just that shyness that pleases.

Although soybean tempeh is the best known, wheat tempeh is unusual and also very good. Make sure that the grain has not been treated with insecticides or fungicide.

Wheat Tempeh

To serve four people, wash 1½ cup (375 ml) cracked hard, red spring wheat in cold water, rinse, put wheat into a large pan, cover with water, soak for 30 minutes and boil for 12 minutes. Drain through a sieve, cool to room temperature, and dry with a towel by blotting. Drying is a very important step.

The mold *Rhizopus* has been used to ferment soybeans and grains into tempehs in the Far East, and contemporary scientists have developed a culture *Rhizopus oligosporus* NRRL 2710 that is especially satisfactory in making wheat or bean tempeh. Prepare the *Rhizopus* inoculant by mixing one teaspoon (5 ml) with four teaspoons (20 ml) water. If you use self-saved starter, no water is needed, but some people advise that two tablespoons (30 ml) vinegar be mixed into the well-dried, precooked grain with the self-saved tempeh starter. Whether using the moistened inoculant or the self-saved starter and vinegar, mix the culture into the wheat thoroughly. When the mold spores are well integrated with the grain, spread the tempeh ½ inch (1.3 cm) deep in several plastic bags that have been perforated with small air holes. Tie the bags and lay the flat cakes on racks so that their bottoms can get air. The mold inoculant needs air to grow. Cover and set the racks in a warm place, 86°F (30°C) for about 26 to 36 hours. The grain cakes will generate heat, but they should not be allowed to get too hot. The plastic will begin to sweat and the mold will begin to grow; when it is white, cottony and marbled with dark gray, you know that your tempeh is coming down the home stretch. Do not eat it too soon because the flavor develops during the last few hours of fermentation.

In good tempeh the grains will be solidly bound together and the cake may be easily sliced. Fry in a skim of oil for about five minutes on each side. Tempeh browns nicely, and it makes excellent finger food for snacks or as a unique, low-salt food to be eaten with salads. The other evening we cooked tempeh fingers in the pan in which oysters had been fried; they were tasty.

Like other microorganisms, molds are greedy; they overpopulate and will die in their own juices. Stopping the action is as important as starting it. To stop the mold's action, you can eat the tempeh immediately, freeze, brine, or refrigerate it. Freeze tempeh for keeping by parboiling it five minutes before packaging; allow it to thaw for 45 minutes before using. Brine tempeh in a solution of one tablespoon (15 ml) salt to ¾ cup (175 ml) water for longer keeping. Fresh tempeh should be refrigerated; if kept at room temperature for more than 48 hours it becomes off-tasting and will destroy itself.

Tofu, Sufu and Other Soybean Foods

Tofu, also called soybean curd, is the basis of a Chinese soybean milk cheese called Sufu. Tofu is an unfermented curd made from soaked, ground, boiled and strained soybeans. The liquid is saved. I met a lady who tried to make this curd but she failed because she threw out the milk and saved the bean sediments. That sounded like something I would do. To make tofu, the saved soy milk is coagulated with calcium sulfate, pressed in a cheese press and the firm curd is immersed in cold water to remove the beany flavor. Tofu is stored in distilled water and when needed it may be sliced or cubed and briefly cooked with Chinese foods.

Sufu is a fermented soy milk cheese made from squares of tofu. The squares are inoculated with molds, brined for a month and matured in a brine/rice wine mixture for two months. This cream cheese-textured, winey curd is an excellent snack and meat substitute, and due to its high salt and alcohol content, it does not readily spoil.

Though the Chinese may not have had stillrooms as such, they habitually tucked away a few jugs of fermented foods against the threat of famine.

In the contemporary Western world, salted soybean curds are made into numerous finger foods and imitation meat products. An example is imitation bacon. Tofu is spun into spaghetti, balled like yarn, immersed in a fermented flavoring agent, colored, then rolled to imitate sliced bacon.

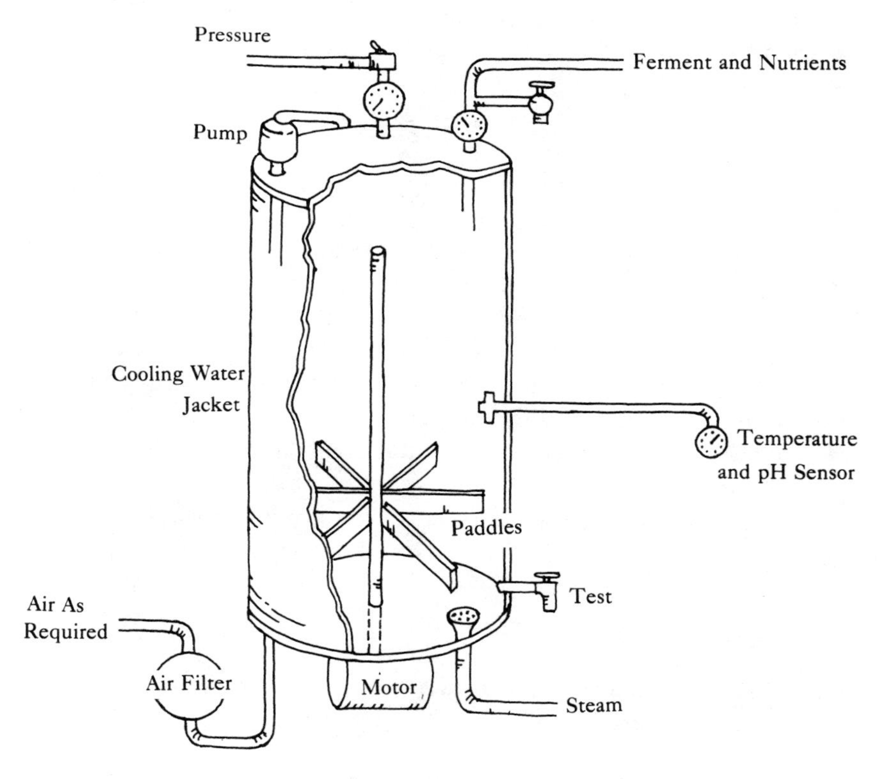

A COMMERCIAL FERMENTER

Used to grow yeast, bacterium for sausage, cheese starters and pharmaceutical antibiotics. The cylinder is filled with water and air and pH are controlled; the growing medium is agitated and the fermented product is harvested when mature.

Soybean foods are a recognized source of low-cost protein. They are earth-thrifty in the sense of earth-surface usage as compared with the land required to grow meat. However, many food producers give too little thought to the destruction that soybeans and other row crops render to the earth's fragile skin when planted on hillsides. There is one truth that should loom paramount in the minds of all man; in the growing of foods, man must protect the surface of the earth. We have only to look at the gashed hillsides adjoining the Great Wall of China, or the deadly expanse of treeless spoil in parts of the Middle East, the United States and Africa to comprehend what a self satisfied pimple mankind has been. To allow our planet's growing power to be ravaged by thoughtless men is to destroy our children's children. Population growth predicts a fantastic need for more foods by the year 2000; while producing that food, man, all men, must look after the health of the earth.

Microbiological Future Foods

Scientists tell us that future foods highlight two bright spots. They tell us that factory-laboratory proteins extracted from wood pulp, paper and plant wastes, and the single-cell protein *Methylophilus methylotrophus* a bacterium grown on methanol, can be eaten. Named Pruteen by West German and British producers, the bacteria oxidize methane and employ it as a source of carbon and energy. The nutritious, light-brown granules are made by growing and drying large numbers of the bacteria. A British factory can produce 75,000 tons a year, but due to petroleum costs, single-cell protein foods have difficulty competing in the marketplace. The Soviet Union has shown that without the cost restraints imposed by having to buy petroleum, these future foods can be produced as a cheap source of nutrients. The drawback is that people must be educated to accept them. After they become accepted, these itty bitty microbes might one day feed the world and save the earth's skin, save her trees' lungs and her heart's blood, which is our planet's sparkling fresh water. Microbes may yet lead the way to planet earth's survival.

Bacteria in Sausage Making

History shows that as agricultural techniques improved, populations increased and the need for preserved food grew proportionately because man clustered in cities. Drying food was most likely the earliest form of preserving, and drying meat probably led the way. One ancient dried meat was sausage. When Roman soldiers went off to do battle with barbarians, they carried chopped and spiced meat that had been salted and packed into gut casings. They employed butchering squads, who regularly ground, spiced, salted and stuffed the army's sausage. These meat men probably did not understand that the lactic acid bacteria in their drying rooms helped to ferment, flavor and preserve their sausages, but they probably did realize that the food value of sausage is similar to that of fresh meat. Fermenting foods generally does not change their nutritional value.

Meat juices are ideal media for growth of microorganisms; they are moist, they contain nutrients and sugar, and they have a nearly neutral pH. As with other food preserving, acids must be introduced quickly in order to deter bad microbe infestation. The conditions of the butchering area determine the microbe activity. Cleanliness, temperature, the presence of lactic acid bacteria: all contribute to the successful sausage.

Fermented sausage falls into two general kinds: semidry and dry-hard. Both keep well in a cool, dry place. Semidry sausage is minced meat, fat, salt, spices and saltpeter, mixed and stuffed into casings. It is then smoked with a cool smoke for a relatively long time before being heated briefly to kill any bad microbes prior to being air-dried for a short time. Though there is no fast rule, dry-hard sausages are spicier; the minced meat, flavoring, salt and nitrates are mixed and allowed to briefly ferment on a table at room temperature before being cased, hot smoked for a short time and air dried for a relatively long time.

In naturally fermented sausages, lactic acid bacteria are responsible for the tang. Though variation in the smokehouse and maturing temperatures play an important part in affecting the final acidity of the product, fermented sausage acidity is generally about 1% and the pH is between 4.5 and 4.8. This acid content deters other bacteria and preserves and flavors the meat.

Pelham's Hard Sausage

Makes 2 four-to five-pound links.

Attach the meat grinder with a fine plate and hitch it up to a strong arm.

In a large dishpan place 12 pounds (6 kg) of lean, unfrozen, preground fresh pork and sprinkle with 12 cloves of minced garlic, ¾ cup (175 ml) salt, five teaspoons (25 ml) crushed or ground red pepper, three teaspoons (15 ml) powdered allspice and one teaspoon (5 ml) saltpeter. Mix thoroughly and feed into the grinder. Catch the meat in a second container.

Add three pounds (1.5 kg) finely diced pork fat, six teaspoons (30 ml) whole black peppercorns, six tablespoons (90 ml) gin, vodka or brandy. Work the meat with your hands, squeezing and mixing the ingredients for about 15 minutes. You will have to stop several times for a sip of brandy during this process because your hands will become stiff.

With one person holding the casing, spoon the sausage into it and push down tightly with the back of the spoon. Fill and firm, repeat until you have used all your meat. (I use an unbleached muslin sleeve made by sewing an arm-length casing.) Lightly salt the outside of each sausage casing, tie and hang the sausage in a cool room, about 60°F (15°C) for about four days to ferment. Put a paper under the sausage to catch the drip, and watch the dog.

Then hang the sausage in a low-humidity, varmint- and fly-free place at about 50°F (10°C) for about four or five months. The sausage will shrink some.

This makes a tasty salami-type sausage that keeps well. I have kept hanging sausage over a year in my screened sausage cage in the cellar; however, they become harder and saltier with age. I have learned that for best flavor, homemade sausage should be eaten or frozen after about five months of drying. I thrill at the sight of sausages hanging in a row; they are instant food, and like a plain dress, they can be gussied up, kept simple, or they may serve to cover a multitude of sins.

My Missouri landlady used to say that sausage is a way of educating shiftless scraps of meat into becoming respectable food. "Would that shiftless people could be pounded into respectability as easily," she added, nodding toward a shack clinging to the bank of a nearby river.

Pelham Smith, a local ne'er-do-well, lived in the shack; he was the hog killer of the community. He was also the sausage maker. At butchering time people gave him meat scraps, and by March he would deliver dried sausages that he traded for cash. Everyone had a good word for Pelham's sausages, but they treated him with contempt.

I had to pass his house on the way to White Dove, the one-room school where I taught. On sunny afternoons I would find Pelham sitting on a bench and I would visit a minute. His face was waxy, half-transparent like milk glass; wrinkle folded upon wrinkle along his neck, and his white hair was tufted like a bunny tail. Beneath his jutting brow, the man's blue eyes gazed audaciously, a contrast to his obviously beaten spirit. The odor of liquor generally surrounded him. Sometimes Pelham uttered not a single word, at other times he would break into a laugh, lick his lips and pass his sleeve across his mouth. Often he propped himself with both palms and swayed side to side. But most afternoons we spoke of the weather, crops or the state of the muddy road.

When I asked Pelham about sausage making, he brightened and invited me into his incredibly cluttered home. Picking my way across the crammed room, I was astonished to see an enormous gray goose crouched amid the rubbish. I stopped.

"Clifford," Pelham chuckled not without affection, adding, "He is not allowed to go out because of them," he indicated up-hill with a toss of his head. I knew that my landlady had nasty-tempered white geese that regularly ran me halfway to school; in time I understood that Clifford had seduced some of his neighbor's geese and that Pelham had been ordered to keep his bird locked inside. One could see that the man was proud of his magnificent pet, and when I admired the gander, he said, "God granted him a virtue; he can forgive. They cannot." Again he nodded up-hill. I did not understand, and looking into his face I saw that his simpleton character had disappeared.

Ill at ease, I reminded him about my sausage-making interest, and he directed me to a winding staircase that led to a white-washed stone cellar with a butcher block, hand grinder, knives, saws, all neatly placed as in a meat shop. The room was divided in half by a walk-in screened area in which sausages hung like balusters. I had never seen such an array of skinny, long, fat and stubby sausages. "You understand that sausage must earn its reputation; it must practice patience. To save itself from the sin of becoming plain salty meat, it must grow in depth and understanding; sausage must ferment," he told me. I asked, if he was a butcher, why he did not open a shop. With a deep sigh he answered, "To have gone wrong is a terrible burden, there is no helping it. I must carry my sin."

His spooky answer nearly catapulted me up the stairwell.

At supper I asked about Pelham and was told I should never, never go into his house. He was an evil man, I was informed. The whole neighborhood knew about him, they watched him, and though he was a first-rate sausage maker, he had given himself up to the devil.

I was to leave him alone, my landlady said.

Being young, that was all I needed to stimulate my interest in Pelham Smith. I went to Mrs. Webster, a palmist and outcast of the community. Though she rolled her eyes in an uneasy way, she said that as a young man Pelham had fallen in love with a "lusterless" girl, the minister's daughter, who possessed a "tragic love line." The couple was to be wed, the woman told me; Pelham was a farmer, brother to my landlady's husband, and by rights should have inherited land. However, before his wedding he was caught "peeking," and the neighbors "took care of him, black-balled him." The preacher moved away, and Pelham "ain't amounted to a crow's caw. No spleen. The only thing he's good for is sticking hogs and sausages," she said.

My heart hurt for Pelham Smith, who had been forced to drain his cup of penance the past 40 years. No animal can be so artistically cruel as man.

Surprisingly, when I next saw Pelham he hastily beckoned for me to come. With a shiver of trepidation I followed. In the creek behind his house Clifford was swimming, and paddling in a circle behind him was a gray goose followed by four tuffs of yellow fluff. "My children," Pelham grinned with eloquent pride. "Patience conquers all things. Like my sausages, suffering, being ground up, so to speak, is not evil; there are devils hidden in all of us. Patience is the thing, patience can make a man worthy and respectable. Clifford gave me a family." And I saw a terrible tenderness creep into his face. Patience and self-sacrifice are virtues of the young in love.

Fermenting food takes patience; working with sausage, cheese, tempeh or pickles takes patience and gives hope. With fermentation there is a tomorrow; a tomorrow for man and a tomorrow for our glad earth.

4

CHAPTER

Flavoring Is a Leader

Flavoring springs from the misty world of enchantment to tremble and mate with man's palate. . . . Flavoring surrenders itself to play before seduction. . . . Flavoring is a solitary meditation, a croix de guerre, a wench without underwear. . . .

A jolly cop from West Virginia, a man who obviously bows to the shrine of flavor, described his taste buds as a "bunch of boys watching flavorful girls go by. A gusty meal, like venison chili, and Whooee! Them skirts fly high!"

Science books say that flavor is the interaction of smell sensations and chemicals that impart a distinctive taste: bitter, salty, sour or sweet. Aromatic flavorings that come from the essences of plants may be alcohols or acids, or they may be combined to form esters. In addition to organic flavorings, seasonings may be created by yeasts or other fermentations, such as bacteria and mold enzymes. Chemistry can duplicate most flavors but natural flavorings are generally preferred because they are honest yet flexible.

Flavoring is a leader; it has led men around the globe.

Flavors of Ancient Greece and Rome

Classical Greeks knew about flavor; they played host to all Oriental, Indian and Egyptian ferments, plus spicy luxuries such as ginger, cloves, cinnamon and black pepper. In addition, the Greeks grew anise, basil, caraway and rosemary; they developed their own exotic vintages and vinegars; they created glorious goat cheeses, smoked tunnyfish and wondrous plump fermented olives. Many of today's keen flavors were first isolated and described in Greece.

As populations of the Adriatic expanded, however, the need to provide more food and fuel with which to cook and heat pushed ancient peoples farther up on the hillsides where, denuded of native trees, the fragile soil bled into the seas. Later wars and the need for money from exports of olive oil and wine persuaded the Greek Golden Age to push its self-destruct button. Through abuse, their land became near-barren, and their population was reduced to near-subsistence living. For centuries savory foods took a back seat to plain eating.

The Roman heyday, though more spectacular as it followed its conquest and payoff road, fizzled in much the same manner as that of Greece. Where the exploitation of the olive and the vine had destroyed verdant Greek hills, and thus their country's strength, overexpansion undermined the goodness of ancient Italy. With illusions of deification, power-hungry Roman emperors bought and sustained their positions by free doles of grain, olive oil and wine to the multitudes. The give-away programs brought allegiance, but state subsidies destroyed the peoples' faith in themselves and howls for public assistance increased. During the third century A.D. the lack of fuel and poor transportation of fresh produce reduced Rome city dwellers to eating faceless food. Even after the Emperor's bureaucratic bakers mass-produced and distributed free bread, the average citizen had little else except a few pickled olives, dried figs, dates, an occasional chunk of pork or fish, plus any flavoring sauce he could liberate.

Zesty sauces became the lifeblood of early Rome. While the common man was chomping state bread, wealthy Romans were munching dishes of roast kid pressed to imitate squab, complete with feathers made of garlic slivers. Their counterfeit birds were served in a bed of goose liver pâté with a mince of Genoa ham, all swimming in butter melissa sauce with mushroom caps and fermented black olives. A second elegant sauce was made of sautéd onions and ginger, black pepper, candied angelica root, the strong asafetida, *Ferula foetida*, also known as Devil's Dung, and capers infused in Madeira wine. The more pungent the sauce, the more the noble palates liked it. Historians state that palace cooks were altered into eunuchs so that they would dream of sauces, not women. Romans also were said to relish the metallic flavor of pewter, an addiction that no doubt galvanized their taste buds so that intensely potent flavors were required to penetrate their armorized gullets. One pickled fish sauce was so popular at the beginning of the Christian era that factories were built in which the anchovy-type fish were layered with salt, allowed to ferment in a warm place before being sealed in containers and buried for a half-year. The flavor of this fish penetrated undercooked pastas and rock-hard bread so effectively that massive 'soup lines' formed to collect the government allotments of sauce.

I tasted fermented fish sauce once. Actually, I fell into its hiding place. One autumn my students and I joined the Quigillingnok village matrons to collect eiderdown caught on bushes after the eider ducks had moulted. We were following the good-natured women, whose bright parkas were blowing in the Bering Sea wind, as they darted this way and that after the white fluff. I stopped. A sort of spiritual quietude descended upon me. The primitive scene, the solitude, the busy motion as far as the eye could see, cast a spell. Telling the youngsters to run ahead, I meandered, leaping from one watery hummock to the next. The unfinished land was reflected in puddles below, clouds scooted, attentive birds wheeled, screaming as they circled. Suddenly I found myself knee-deep in mire and sinking. Quicksand! Remembering books I had read, I contemplated flinging myself full length for bouyancy, but I could not bring myself to lie down in the mud. I began to yell. Never had such unladylike sounds escaped me.

Unfortunately, my mud hole was located in a dip in the landscape, so no one could see me. As one grassy-root handhold after another pulled free and I was being drawn farther and farther away from solid ground, I screamed the name, "Timothy!" Being a new teacher, I could remember only one name, "Timothy!" There was no answer. I could hear natives chattering, calling to each other, but no one missed me. There was nothing to hold to in that slime; I could not lift my heavy feet.

Heavenly angels could look no sweeter than the square-jawed Eskimo lad who momentarily appeared, muttered, "You in the mud," and disappeared.

Slowly I sank down, yelling until in one delicious moment I saw a crowd of wide-eyed Eskimos around me. Clacking in their native tongue, one, two, three, four husky men leaped into my mud bath. Each held onto the other to form a human chain that connected to dry land where a cluster of people tugged at the man closest to shore.

I was glad that I had long arms. The men pulled, now joking with each other, now encouraging me; but when I was almost to safety, they dropped my arms and actually turned their backs on me. It seemed an hour before they again reached for me and I learned what had taken precedence over my rescue. One man had discovered a submerged sealskin bag, which was recognized as containing fermented fish. It had been buried several years earlier. I have since been told that when the frail soil of that boggy land is disturbed, water reclaims the area and seemingly bottomless pits of mire take over.

Though my modesty had been challenged when the cold had compelled me to change clothes on that treeless flat, I had not been hurt. My ego had suffered some, too, because after I had shimmied into a short Eskimo's parka, from which my pale legs extended like a sand crane's, my companions giggled hilariously. I loved it.

The following evening I invited the villagers to celebrate my survival with cookies, crackers and cocoa; plus I gave each a lemon. Those people were addicted to lemons. They named me O'goayook, which means Yellowlegs, a spindly-shanked bird, and as the evening progressed, each rescuer told stories with actions mimicking my sinking, Tim running for help, their jumping into the mud, my loping home; but in every story my quicksand caper was a subplot, so to speak. The finding of the fermented fish was their triumph. Who had buried it? When? Was there more? Periodically the storytellers would take a squeeze of lemon, make wry faces and everyone would laugh. It was a funny-face contest, and as a climax to the evening everyone was given a taste of the fermented fish juice.

Cheesy, musky, exquisite! Each taste bud eagerly begged for one more drop. But no, the one sack would be divided among all village homes where it would flavor food for the winter. Those Eskimos were a giving people, but I discovered that they also had no conception of the clock. Everyone stayed all night. Some nodded momentarily, waked and started storytelling again. When at last they bowed to thank me for showing them the fish cache, I bowed back, and tottering on my feet, I thanked them for sharing their warm hearts and their superb sauce, and incidentally for saving my life.

Pickled Jackfish, as they called the flavoring, was made by roughly chopping cleaned, late-running river fish and layering them with salt and a little polygonum weed in a seal bladder. Several bladders were hung high on the fishracks behind a house until a brine was formed, then the bags were sealed by tying the doubled-over orifice with gut line. They encased the bladder bags in sealskin sacks, hair side out, and buried them into the frost line, heaped over with soil and dog or other droppings in order to discourage wolves. The sacks were customarily left in the ground for a year.

That caviarlike fish sauce exploded with tiny bursts of intense flavor, and like a little saucy flattery, a mite went a long way.

With populations vying for space on our beloved planet, it seems apparent that cookery will tend toward employing those flavorings that give the greatest variety of tastes and nutritional value with the least expenditure of the world's finite resources. Premade sauces to dress up basic foods may become a universal mode of eating.

Some sauces, like an innocent child, are scrupulously naive and seem to enjoy giving pleasure to mundane food. Some are of haute cuisine elegance, flawless in the mingling of their charms. Sauces may insidiously prolong the memory of their goodness like the face of a long-ago friend; some generate deep affection.

Flavorings may be divided into three categories: those based on meat, seafood, fowl or milk products; those that utilize fermented ingredients such as wine, vinegar, soy sauce; and those flavorings created from plant parts and flavorings that utilize natural plant essences or the chemical substitutes of volatile plant oils. Very often the three types of flavoring are combined, such as in duck sauce, which includes juices from the roasted bird, a clove-studded onion, rosemary, wine and distilled orange peeling water.

Saving Flavor at Home

Volatile plant oils to be used in flavoring may be extracted by boiling, or by dissolving the essence in ethyl alcohol, vinegar or wine; or plant flavorings may be recovered by distillation.

Many people regard home distillation as illicit; thoughts of no-no whiskey cluster with ideas about sophisticated medicines and perfumery and scare people. Actually, distillation of flavorful waters is simple and legal. No permit is required for the possession of or the use of either a boiling water or a steam distillation apparatus for making food flavorings.

The key to understanding the distillation of plant essences is the awareness that different substances boil or vaporize at different temperatures. Many volatile plant oils, like alcohol, boil before water;

they are like alcohol, not greasy. Though most flavoring distillates are commercially produced by use of alcohol solvents and the distillation of the flavored alcohol in vacuum stills, a general knowledge of plant oil vaporizing temperatures is important for home distillation of flavoring waters.

I have read about boiling water stovetop stills being fashioned out of pressure cookers, but I am wary of makeshift contraptions when handling hazardous substances such as steam. Manufacturers of laboratory equipment (such as VWR Scientific, Box 999, South Plainfield, N.J. 07080) sell small glass stills and components.

Boiling Water Distillation

Boiling water distillation of plant material generally follows four steps: 1. On a rack in the boiler part of a still, the plant material is placed and water is added. 2. The head of the still is fastened, the condensing tube is secured and a check is made to be sure that the cold water runs freely through the condensing tube jacket. Or if a worm and a barrel is used, the barrel is filled with cold water. 3. Vaporizing temperatures of the materials to be distilled are noted and the still is heated as quickly as possible to just above that temperature. Any steam leaks are sealed with clamps, and a narrow-mouthed collector with a lid is placed under the condensing tube exit. 4. Generally, not more than one-fourth of the original amount of boiler water is collected as it perks through the condenser; the still is closed down at that time, the collector is capped, and the boiler is cooled, washed or recharged with fresh material.

Some people concentrate their herb waters by distilling their distillate a second or third time. Sometimes they pour the first distillate over new material and run the batch a second time. I usually run my herbs once. It is like catching a bus: If those little plant oils are too lazy to run after their steam vehicle, let them stay home and make the compost heap happy. I know people who are too indolent to get out and scramble for a little excitement. They sit home and complain that life is passing them by; generally it is.

I use boiling water distillation for bulky materials; for small herb and spice distillation I hook up my swifty-nifty, backyard steam still. A bright young chemist friend made it for me when he learned that I was distilling everything in my ponderous boiling water pot still. Friend Thomas Birkel is right; steam distillation is the way to go, especially if its condenser is connected to a dribbling garden hose.

A Steam Still for Plant Material

A steam still consists of two flasks that are heated on two fires. I use a camp stove. The larger boiling water flask directs steam to the smaller still flask that contains the material and two tubes inserted in its cork. The steam tube extends nearly to the bottom of the small still. The second stopper insert is a short glass tube that directs vapors from the material being heated to a condenser, which is cooled by running water. Overflow water runs onto the grass and a narrow-top jar is placed under the inner condensing tube to catch the good stuff.

On stilling days I usually connect up my apparatus, light a fire under the boiler flask, run and snatch a couple of handsful of the herb to be distilled, and put about two inches (5 cm) water in the small still together with my herb. I light a second low fire under the still, fasten the corks and tubes in place, and stand back. Actually the two stoppers are permanently fitted with their glass tubing inserts, and all that has to be done is poke down the corks and fit the glass tubes poking out of each cork into their flexible tubing carriers.

As boiler water vaporizes, steam enters the small still through the long steam tube, which intrudes through the material and into the bottom water. The steam heats the water and a watery vapor filters upward through the herb, gleaning oils as it travels to the exit tube. As the first moisture appears at the top of the condenser, it swirls in a silken fashion like alcohol before it gathers strength and drips through the condenser. It makes a clicking sound, not unlike an adolescent cricket that has not quite gotten its act together. Hearing a still sing is an exhilarating experience.

After collecting all of the flavoring desired, I uncork my still stopper, being careful not to shatter the glass tubes poking through it. I shake out the spent herb, rinse the still and run another batch. I collect the watery extract in small dark bottles, lid them and race away to gather more material. Counting setting-up and closing-down, small amounts of eight herbs can be distilled in about three hours. They are finger-tip ready for sauces, salads or schnapps.

"Why bother to distill?" one might ask. Distilled extractions contain a more uniform flavor; they may be measured with ease; when properly handled they do not readily deteriorate. Most druggists will sell small dark bottles, and if you are in doubt about the keeping qualities of your distilled herb waters, weep a tiny bit of vodka down the inside of each bottle so that it floats on top.

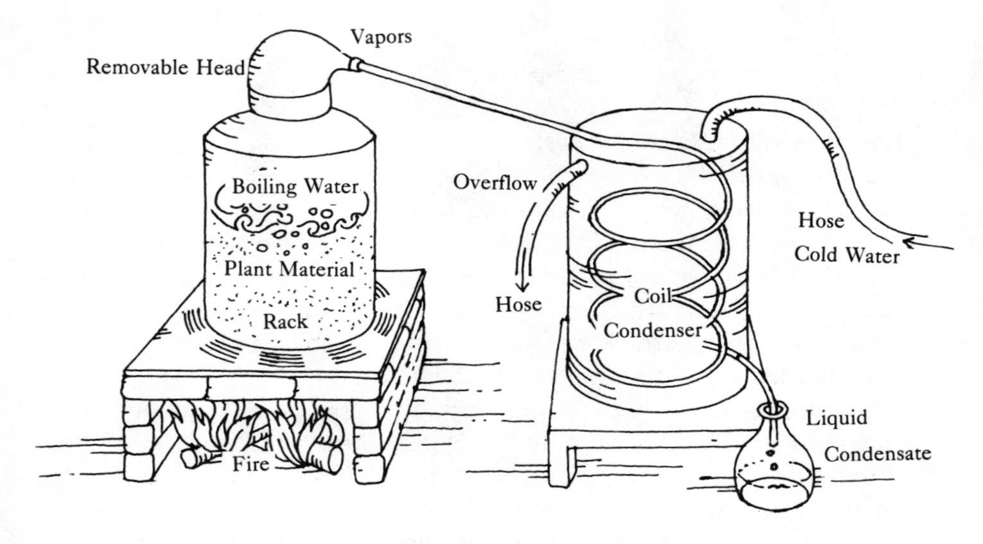

COPPER BOILING WATER STILL WITH COIL CONDENSER

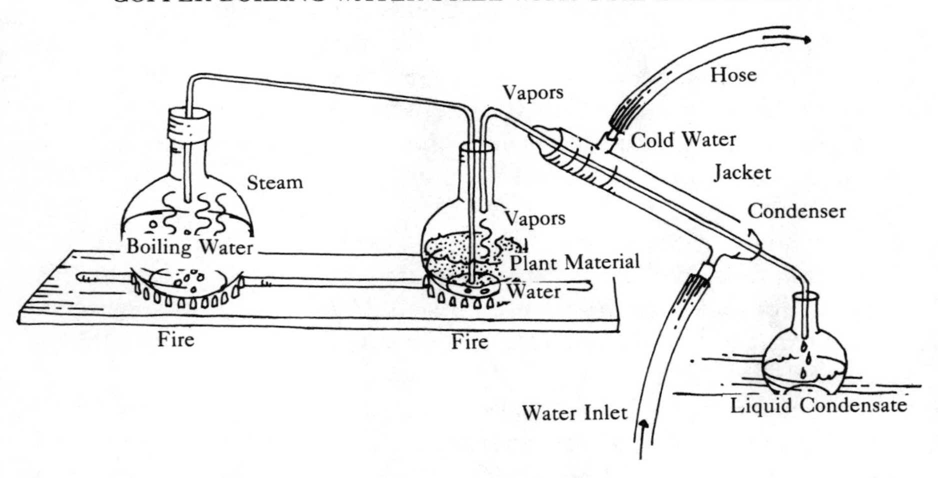

GLASS FLASK STEAM STILL

Drying, Steeping and Synthesizing Herbs

If you cannot distill flavorful water extracts, dry your surplus herbs. Drying herbs is the oldest way of saving flavor; however, some loss of zip occurs in drying. In general it is best to harvest herbs for drying on a sunny day just before they shoot to flower because most plant oils concentrate their oils at that time. Cut herb stalks and hang them loosely in the shade. Do not allow them to get wet, and bring them under cover at night. As soon as leaves are crunchy, (about four days), strip, crumble and put them into small, clean, dry, dark jars with tight lids. Label and store in a cool, dry place.

Some herbalists extract plant flavors by macerating or steeping the crushed herb in alcohol, wine or vinegar. They recharge the same liquid with new material several times until the carrier becomes saturated with the volatile oil. If alcohol is employed, it is important to use potable ethyl alcohol of the correct proof (given here with the description of each herb) and to store the extract in a tightly lidded, dark bottle in a cool place.

During the latter half of the nineteenth century, commercial spice firms began studying methods of extracting volatile plant oils. They discovered that many spices yield as much as 10% of their weight in essential oils, and that flavoring distillates were potentially marketable for use in canned produce such as soups, sauces and, more recently, in frozen foods. Industrial food chemists, aware that plant oils are often soluble in alcohol, extracted spice flavors by the use of solvents. They soaked the spices in the solvent and distilled the liquid using fractional distillation, usually in vacuo, that is, in a vacuum still. The vacuum still required less heat and with the use of fractional distillation equipment, the food chemists were able to take off certain oils at specific temperatures. The essential oils, thus separated, were sold to canners and the solvent was recovered and used again.

By the mid-1920s most flavoring extracts were chemically synthesized from coal and wood pulp. These cheaper, but often inferior, products have threatened the production of natural spices in lesser developed countries where the cultivation of aromatic plants are sustaining industries. Today, flavorings in many homes have turned to garden herbs and natural spices, but industries such as frozen foods and bakeries continue to employ concentrated synthetic essences.

Empiricists believe that a person's taste, like the moon, is constantly changing; it either grows or wanes. During the doldrom days when your taste buds lie limp, remember the moon and help yourself through dark phases with herbs and spices.

Herb Descriptions and Distillation

If you plan to distill your own flavorings, start with what are known as sweet herbs. Actually, sweet herbs are not sweet. Like sweet grandmothers, they possess certain captivating characteristics and may be endowed with zesty roots, lusty seeds or greenery that adds interest to their company. But sweet herbs can be volatile. They can strike out when their polite forebearance is provoked, and sometimes their chemistry is hostile to surroundings. Nobody likes a hostile herb. However, with a little understanding their flavor can be brought out or their essence distilled and preserved for future use.

As with many families, the sweet herb clan is torn by a rivalry between its two great houses. The *Labiatae* family, best known for its leaves, and the *Umbelliferae*, known for its seeds, cannot abide each other in the same dish.

Labiatae cousins are commonly called mints: the horsemints and horehounds, are the street fighters of the tribe. Basil, marjoram, peppermints, rosemary, savory, sage and thyme carry the highborn oils of this family. Most mints are the rollicking, rough-mannered, generous givers of the plant kingdom that could be likened to the good-natured immigrant Irish grandmothers of yore. They thrive in temperate America.

The *Umbelliferae*, on the other hand, come from a long line of Near-East aristocrats. There is nothing common about these uppity compounded umbels. Anise, caraway, coriander, cumin, dill and fennel traditionally have flavored cheeses, pickles, cakes, liqueurs and sausages. Miserly of leaf and hollow of stalk, the *Umbelliferae* aromatic oils are generally located in ducts near their fruits. Though a few species of this clan have naturalized themselves, many must be coaxed to live in new climes.

In addition to the mints and umbels, other flavorings that have teased palates to insure a seat on seasoning shelves include: capsicum pepper, cardamoms, cinnamon, cloves, ginger, mustard, black pepper, tarragon and turmeric. Root, fruit, herb or distilled water, they all give new life to languid dishes. Flavor leads man around by his taste buds just as surely as a statesman leads a nation.

The Labiatae

Basil, *Ocimum basilicum*, is perhaps the most provocative mint around. Hailed by the Byzantine Greeks as a holy plant that was said to have spontaneously sprouted in Jerusalem in memory of Christ's goodness, the plant has not been without some morbid encounters. Scorpions were said to hide in the herb while waiting to invade human brains. In literature the basil pot has been the recipient of several severed heads. Conversely, in Italy, it was with a merry heart that young suitors showed their honorable intent by wearing basil behind their ears when they went a-courting. I wonder if the gentleman mentioned in Harvard graffiti, "She offered her honor, he honored her offer. . . ." wore a sprig of basil.

Most basils are touchy plants: They exhale a spicy aroma if only slightly bruised, and they are extremely sensitive to the cold. I had a grand aunt like that; she was thin-skinned, bruised easily and always had cold feet.

Contemporary Italian cooks can hardly cook without pesto sauce, made by blending basil with olive oil, garlic, parmesan cheese and minced nuts. Tiny steamed shrimp that have been marinated in basil water, lemon juice, olive oil and crushed garlic, then stuffed into topless cherry tomatoes, create perfect finger food.

Basil distills into a fragrant flavoring water. Place the chopped basil herb on a rack in the still and add a quarter as much water. Heat the still briskly at first, but as soon as distillates begin to vaporize, lower the flame. Sweet basil volatilizes between 113° and 215°F (45° and 101°C), and it seems to run its course very quickly, so the normal rule-of-thumb—distill ¼ of the original liquid—does not hold with basil. Instead, taste and close down the still when the distillate begins to lose its strength. Basil oils are said to be soluble in 160 proof alcohol, and commercial firms that sell the extract to sausage makers and salad dressing companies recover the flavoring by distilling alcohol-basil infusions.

Marjoram, *Origanum majorana*, a symbol of bliss, was given to ancient Roman brides to tuck into their bosoms. The herb was also strewn in outhouses, whether as an air freshener or for bliss-on-the-potty, the old tales did not tell, but they do suggest that marjoram keeps snakes away.

We have a patch of wild marjoram, *O. vulgare*, growing near the bear mud wallow. Several times a year when I need the fresh herb for pizza sauce (tomatoes, garlic, cumin, hot peppers or chili powder, salt, sugar and a fistful of marjoram), I wander down to the patch rattling a bucket of rocks as I go to tell Mr. Bear I'm raiding his marjoram. Lewis told me that if I see him I should sprinkle some marjoram on his tail. Not me. The only tail I would be interested in would be my own flying out of the woods.

Marjoram begins to volatilize at about 163°F (72°C), and the yellowish extraction, which is completed at about the boiling point of water, will take the skin off your tongue. I have not used it for fear of ruining our perfectly happy gullets.

Mints, *Mentha piperita* or *M. spicata*, the oils of which contain 17 different substances, are like cops with years of street time; multitudes of different experiences give understanding. There is no end to the variations that mints give to foods. Mint has been popular in Middle East lentil-onion soups, in purées of eggplant and cheese, and as a dressing with cucumber, chive, olive oil and sliced orange salad. I have sprinkled mint on all manner of green salads and never cease to be surprised by its crisp, pleasant taste.

Mint herb should be wilted for a half day or so before distillation. With boiling water, the volatile oils begin to break down at about 200°F (93°C), and they become highly vaporized at 215°F (101°C). Most mint oils are soluble in 140 proof alcohol.

Rosemary, *Rosmarinus officinalis*, the plant that Greeks braided in their hair to stimulate their thinkers, has never done a thing for my gray matter. I murdered my neighbor's plant that had been left in my care. It simply turned its head to the wall and died; nevertheless, I have been trying to make up to rosemary. I tried it with pork roast but the gravy lost all of its rich meat flavor. I tried it to spark a dead beer—the result was deadly. Rosemary also overpowers my soups and chicken dressing. I think it is vindictive. Rosemary is said to distill readily with vaporization beginning before water boils; its oil yield is said to be eight ounces by weight (225 g) from 100 pounds (50 kg) of flowering tops, and it is said to be soluble in 160 proof alcohol. I have not distilled rosemary; we have enough trouble with our relationship. Maybe someday it will forgive me.

Sage, *Salvia officinalis*, reminds me of a loudly argumentative, though not coarse, teaching companion in Alaska, who was to be pitied but who regularly got my goat. We shared a renovated Army tempo building; her classroom was located in one half and mine in the other. Our mutual cloakroom, a lean-to, ran the length of the building. I am a seether, not a fighter, and the day that I looked outside to see the snow littered with coats, boots and hats, I seethed plenty. Some of my students had hung their clothing in her half of the cloakroom and she had thrown it out. I put on my boots and picked up the garments with no word, but the heat of my seething melted a goodly path through the snowbank. The next day the woman arrived in our classroom carrying two huge Yorkshire puddings and a bucket of grilled sage sausage patties. I never understood her.

Since then sage reminds me that I should have tried harder to have understood her. She died aboard the ship *Yukon* when the steamer split in half in the North Pacific. She was going home, retiring after 30 years of teaching; she never made it. Like many strong-minded people who feel powerless and call attention to their plight by using the only weapons they possess, voices and audacious acts, sage requires understanding; restraint is needed.

Sage distills readily, vaporizing in the still at about 150°F (65°C), and continues coming over until just before water boils. Oils from crushed leaves are soluble in 140 proof alcohol.

Sage dries with abandon, and a plant may be brought indoors so that you may have a fresh, meaningful relationship with it all winter. We enjoy sage in dressing and with cream cheese in snacks; it is somehow reminiscent of Roman antiquity when soldiers, remembering tales of its life-saving qualities, chopped it into their porridge. For us, sage finds its happiest home in fresh pork patties.

Pork Patties

With your hands, thoroughly mix ten pounds (5 kg) of twice-ground pork comprised of ¼ fat, with scant ¼ cup (50 ml) salt, two tablespoons (30 ml) powdered sage or sage water, and two tablespoons (30 ml) black pepper. Form into patties, separate with wax paper and freeze, or fry a big mess of patties for supper. Do not forget to make a bowl of all-American cream gravy with the drippings and serve with grits.

Savory, *Satureia hortensis*, like its brother *S. montana*, seems to climb over into the condensing tube at a whisper of steam from the still. Fresh savory or distilled savory water dresses up peas and asparagus. It's sprightly in stuffing and just great in pork pie. Most vegetables seem to delight in a swim of buttery white sauce sharpened with a dash of savory and grapefruit or lemon juice. Most of the pungent oils from the *Satureia* family pass over into the condenser at about 176°F (80°C), and they are reported to be soluble in 160 proof alcohol.

The Umbelliferae

As my grandmother used to tell me when she mixed mints into sausage, sauces or cooling drinks, "Mints are as free of suspicion as an unlaid egg." Not so with some umbel family members. Many *Umbelliferae* look-alikes are deadly poison: water hemlock, cow bane, fools parsley, poison hemlock. The caution is to clearly identify before harvest. Distilled poisons can kill just as dead as fresh ones.

Anise, *Pimpinella anisum*, also known as aniseed, was so popular an *Umbelliferae* in England during medieval times that it was said to have been taxed to raise money to build the London Bridge. During Rome's days of glory, a spiced cake was presented at the end of a feast; the anise, cumin, cinnamon, clove and orange-flavored cake was thought to prevent indigestion; it was also believed that anise cake would deter the Evil Eye from eyeballing diners.

Licorice-flavored anise seeds, which contain nearly 20% protein and a large amount of fat, contribute greatly to confections. Some people enjoy anise in white sauce and in chicken and rice dishes. We enjoy anise with cooked pears and love it in hard sausage. Anise seeds should be crushed before distillation, and though some vapors pass over into the condenser earlier, I found that anise volatilizes very close to the boiling point of water. Oil seems to congeal on the top of anise water distillates. I have employed anise distillate in cookies. Crushed anise seed is soluble in 180 proof alcohol.

Caraway, *Carum carvi*, is believed to be one of the earliest flavorings cultivated. A seed carried in one's pocket chases away baldness and attracts money, Mr. George, a Missouri herb man, told me with a wink. He also said that he fed his spent caraway plants to his cow so that she would give milk that tasted like rye bread and butter. If you ever find yourself caraway seed rich (or dill rich), put some seeds into a sauce dish and dip crunchy apple slices alternately with cheese cubes for an after-supper nibble.

Superstition surrounds caraway seeds. People of yesteryear believed that if you had caraway in your house, garden or around the chicken coop, no one would steal from you. Caraway was said to keep a roaming husband home and tame wild women. I can attest to none of these beliefs.

Caraway seed must be crushed immediately before distillation. I have had stubborn caraway; some vaporize early, but others seem to wait until the last dog is dead to leave their seedy homes. The literature states that caraway volatilizes between 195° and 240°F (90° and 115°C). Caraway water has been used in sauces where whole seeds may be offensive. Traditionally, extracted caraway oil has been used in Germany to make kümmel. I did not recover enough flavorful distilled water to make distillation of caraway seeds worthwhile. Besides, we love the seeds in rye bread, and they are said to contain 20% protein.

Caraway Candy Comfits

These old-timey sweets are made similarly to candied mint leaves, candied violets or rose petals. Slightly beat an eggwhite, mix in enough caraway seeds to moisten and drop the sticky seeds by spoonfuls onto wax paper on which white sugar has been sprinkled. Coat with sugar, and dry the comfits overnight.

Coriander, *Coriandrum sativum*, derived from the word *koris* ("bed-bug") for its odor, has a peppery leaf, which is used in bean salads and tomato dishes. Coriander, said to be indigenous to the Middle East and Egypt, was introduced into Europe by the Romans. To this day distillers employ coriander seeds in making gin, and bakers mix the peppery-sweet distillate into dark breads.

Known in Mexico as *culantro* and in Chinese markets, as heu-sui, the herb is believed to give special virtues of longevity and virility. The herb grows readily but does not dry well. Collect seeds when dew sticks the easily-shattered seeds to their umbel, place some of the herb, root attached, in a plastic bag in the ice box for fresh greenery in the fall, and freeze some leaves.

Puréed seeds blended with nuts, garlic and olive oil create a tasty sauce for bulgur wheat, rice or tempeh. Indian curry, corned beef, and eggplant dishes rely heavily on coriander extracts for flavoring. Try the tiniest bit in chicken soup.

To my senses, when coriander seeds are distilled, the stimulating liquid smells like hot dogs. Coriander extract is distilled from crushed fruit, and it comes over fast, most oils vaporizing at around 180°F (82°C) by steam distillation. The oil is said to be soluble in 140 proof alcohol.

Cumin, *Cuminum cyminum*, is a shy Middle East plant that grows in Virginia, but I have never distilled its seeds because my plants must have been neutered. They had no seeds. This heavily-flavored *Umbelliferae* was once considered a medicinal, but times have changed, and cumin seed, which grows in umbels like its cousins dill and caraway, has gained popularity in Chinese sauces, South American chili dishes and in Indian curries.

Even though a cumin plant was said to keep hens from hiding their nests, Grandma never planted cumin in Missouri because she said that she did not want her eggs to taste like hot tamales. I loved tamales, and whenever I could maneuver a nickel I would race for the Hot Tamale cart. After unpeeling the corn husk, I would eagerly sink my teeth into the tasty, mellow mush. I vividly recall the peddler with his sad, expressive eyes, hair that stood up like a brush and his rich brown skin. He was said to be a Mexican who had secretly carried off his love from her father's house. The day after the birth of their child, his wife's father killed his daughter, stole the child, and forced the Tamale man to leave his homeland. As I ate I used to wonder at the peddler's solemn eyes and the secret sorrow that shadowed them.

Curry Powder

In a heavy skillet toast two tablespoons (30 ml) each poppy and mustard seeds. When they begin to pop, remove skillet from heat. Mince one small dried hot pepper and add to the skillet together with two tablespoons (30 ml) cumin and one tablespoon (15 ml) each ground coriander and ginger. Stir thoroughly. When well mixed add two tablespoons (30 ml) ground turmeric and bottle in a dark, well-lidded container. This is pretty potent.

Recently, at an EPM Publications office party, a young editor brought a dish of hors d'oeuvres that literally glowed golden. "Curried Cauliflower," she announced shyly as everyone crowded forward for a nibble of the pretty snack. Spicy, a bite of pungent pepper, the warm depth of coriander, her easy-to-eat finger food was as good as it looked.

"How'd you make it?" "Simple," she said in her quiet way. "Cut one head of cauliflower into pieces and simmer until tender in water, a little milk, a pinch of salt, if desired, and two tablespoons of curry powder.

"Drain, chill and marinate four hours in French vinaigrette." (A personal variation of vinegar, a little water, a spoonful or so of sugar and oil.)

I tried the recipe on Lewis who is no vegetable fan and is not in love with curry, either. "This is good!" he exclaimed. I now have a zealous curry convert on my hands as he urges me to try this curry marinade on all kinds of garden goodies.

Dill, *Anethum graveolens*, from the ancient Norse word *dilla*, meaning "to lull," is said to have escaped the northland and spread throughout the world. During medieval times it was reputed to be an aphrodisiac. There is a story involving a Roman goddess, the beautiful, fleet-footed Phyllisa, who abandoned her espoused on her wedding day. The groom, a young racer from the northern mountains, gave chase and, veils flying, caught and seduced Phyllisa in a dill patch. Subdued, the bride emerged with golden flowers in her hair. The dill blooms had turned from white to the gold of gladness, villagers said. Thereafter the enchanted weed was seen as a sign of submission.

Submissive or lulling, dill has captivated taste buds ever since. Every home with a sunny spot should grow a plot of dill. Its feathery leaves are splendid in salads, sauces and spreads. Dill seed heads, as with other *Umbelliferae*, must be cut just as they begin to brown because they quickly shatter. Dry the heads, strip the seeds and store in a dark, well-lidded container. Crushed seed is generally distilled by steam because the volatile parts do not vaporize until about 240°F (115°C). Commercial dill distillates, made by macerating the fruit in 160 proof alcohol and extracting in vacuo, are employed in pickles and sausage. I tried my watery distillate in bread but it added nothing. Dill water used to be given to infants, and a few drops of dill oil dropped on a sugar cube was given to young children while a mother was fixing supper. It was said to nourish and quiet the youngsters so that they would take food slowly and not generate gas. Dill leaves, when chopped very small and stuffed into a bottle before filling with vinegar, contribute a toothsome tone to salad dressing.

Fennel is fickle. This leggy *Foeniculum vulgare*, may also be promiscuous. It crosses with other *Umbelliferae* to produce muddy-flavored offspring. In addition its name is said to mean phoney, an outgrowth of ridicule toward churchmen's habits of covertly eating fennel seeds during their long services. The lightly licorice flavored herb is used to flavor white sauce.

Crushed fennel seed is commercially distilled in vacuo. Books indicate that its constituents, which include anethole, the chief oil of anise, vaporize between 158° and 257°F (70° and 125°C) and are soluble in 180 proof alcohol. I have not distilled it because I have never been able to get the herb together with my still. Fennel grows wild along the Eastern Shore of Virginia, and even on a gray day, when fennel is in bloom, the world shimmers as if smiling in sunshine. It was an old practice to boil fennel with fish in order to cut oiliness. Virginia shore fennel may have escaped from kitchen gardens where it had been planted for that purpose.

Parsley, the stately archbishop of herbs, lends graciousness, nutrients and absolution to the world. It is forgiving; parsley's value as a breath sweetener attests to that. Whether it be whiskey, garlic or onion, this pretty, nibble and salad sprite is the perfect odor-eliminator. Parsley does not distill well, neither do the *Umbelliferae* cousins lovage or angelica; they simply refused to be captured by my still.

Flavored Vinegars

Since Roman times, bruised members of the *Labiatae* and *Umbelliferae* families have been infused in vinegar or wine, giving their essences to the liquid. Natural vinegar is the product of dual fermentation; first yeasts ferment sugary liquids into alcohol. Then airborne microbes, the *Acetobacter aceti*, or vinegar-making bacteria, oxidize the alcohol.

My grandfather made vinegar by filling the wine barrel he labeled *Vinaigre* ⅔ full of bad wine, leaving the bung hole open and allowing a jellylike substance, called a mother, to acetify his wine. It used to be believed that the mother on vinegar was caused by a small living fungus called *Mycoderma aceti*, but scientists have established that vinegar is produced by bacterial fermentation. Granddaddy kept his vinegar keg in a separate place because he did not want his good wines to become infested with the *aceti*. Natural vinegar should contain 4% to 6% acetic acid, and it may be diluted with distilled water if it becomes too strong through dehydration. If you have some natural wine that has not been made sterile by chemicals, get a few chips of wood, soak them in natural vinegar and poke them into the wine bottle. Do not lid tightly because healthy vinegar-making microbes need oxygen. They also like temperatures to be about 86°F (30°C).

In addition to sweet herb vinegars, infusion of slightly manic spices and garlic create flavorings that are especially useful to salt-restricted dieters. Flavored vinegars were popular in Paris during the Middle Ages. Children pulling wagons from door to door, called out "Vinaigres bons et biaux, du bon vinaigre!" as they held out pitchers of their wares. Tarragon, mustard and garlic, and shallot vinegars were popular in those days. Wine vinegar is tart and acid in flavor, and it retains an aroma of wine. Cider vinegar is topaz, it is less acid than wine vinegar. Beer vinegar is also yellowish and has a slightly bitter taste. Wood vinegar is clear, generally called distilled, and it has a marked acid flavor.

Other Spicery Distillates

Like other insecure entities, the mint herb tribe and the umbel clan have tried to dominate the flavoring world, but there have been revolts. Revolutions generally follow revolts, and during alternate centuries the exotic spices, allspice, capsicum, cardamom, cinnamon, clove, garlic, ginger, mustard, pepper, tarragon, plus a whole flock of nutty seasonings, have marched like warriors against their herbal adversaries. Flavor should live in harmony; certainly there are enough gullets to go around, but herbs are stubborn. It takes strength to lead.

Allspice, *Pimenta dioica* is from a Jamaican tree, whose unripe berries are sundried and used as a flavoring reminiscent of cloves and pepper. Columbus and his chums were said to have introduced allspice (then wrongly called pimento) to spice-hungry Europe, where it gained popularity in pickling and in steamed puddings.

Using a small boiling water still, immerse one part freshly crushed allspice berries in two parts water and boil gently until one-half of the original liquid comes over through the condensing tube. The agreeable constituents volatilize between 122° and 203°F (50° and 95°C), so the still water should be just simmering. Most allspice oils may be dissolved in 140 proof alcohol. Pimento water may be used in soups, sauces, meatloaf and pies. It is also savory in a punch made with lemon, tea, Angostura bitters and ginger ale. For young people, sweeten the beverage and top with sprigs of mint and a violet. The commercial oil of pimento is employed to add a spicy note to Chartreuse liqueur and to some Benedictines.

Anise or **Star Anise**, *Illicium verum*, of the magnolia family, no relation to the milder *Umbelliferae* aniseed, is an ancient tree of western Asia that yields commercial anise oil distilled from its crushed star anise fruit. Sixteen-year-old trees, harvested three times a year, yield about 60 pounds (30 kg) of seed.

A Chinese star anise still boiler consists of a wooden vat, the bottom of which is perforated and rests on an iron kettle that is supported by a masonry fireplace over a fire. The still is charged with crushed star anise by lifting the head, and when closed, the vapors coming from the water boiling below the perforated boiler bottom pass through the aniseed into the condenser, which is a glazed, earthenware vessel sitting on top of the head. The condenser is fitted with a concave kettlelike cover into which cold water is passed by means of a bamboo tube. After the contents of the boiler are heated, the water vapors, saturated with oil, pass upward into the condenser where they hit the cold kettle top. The distillate, as it drops, flows into a tube placed at the base of the earthenware condenser and is conducted away from the still to a collector. Large stills, holding 360 pounds (180 kg) of seed, require about 48 hours of firing and they yield about 3% star anise oil that is highly sensitive to light and air.

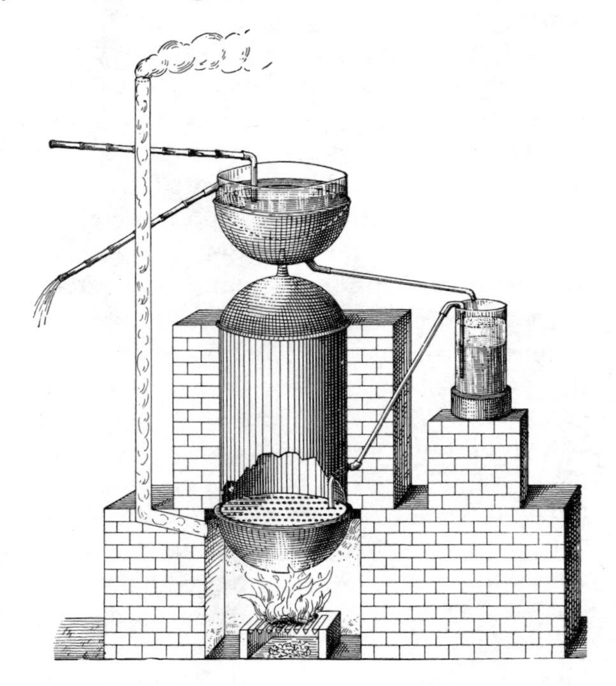

STAR ANISE STILL

Capsicum, *Frutescens*, hot peppers, are like women. Some are enchanting to look at, temptingly warm, fleshy and mellow; some are sublimely lazy and unbelievably succulent; others are pure dynamite. Unfortunately, like many self-centered beauties, hot peppers have not learned to live in harmony with other foods, and they must be handled with care. Fruits of the Central American capsicum plants, which are cousins to the tomato and nightshade families, circled the globe with Vasco da Gama. All tropical peoples embraced them, planted them and recrossed them to produce Oriental, Indian and African peppers. The Aztecs ate a maize porridge sweetened with honey and flavored with hot peppers; later Mexicans made maize tortillas and ate them with beans, tomatoes and hot peppers. Asians created sweet, hot sauces with onions, garlic, cardamom and capsicum peppers as a flavoring for their grain pastes and pancakes. Middle East people blended hot pepper into sauces made with fermented fruits, cardamom and olive oil to be eaten with their fritters and bread. East Indians chopped hot peppers with ginger and garlic, coriander and cumin in ghi butter as a seasoning for their wheat and rice.

Soluble in 160 proof alcohol, the oleoresins of *capsicum* peppers are commercially distilled for use in beverages and sausages. Hot sauce has been made by fermenting ground hot peppers into a fiery wine, turning it with the addition of *aceti* bacteria to form a vinegar product before aging, filtering and bottling. I have fermented hot pepper seeds and membranes left over from cutting the fleshy parts for pepper jam. The pepper seed's umbilical cord holds its pungency. Later I tried to distill the luscious red liquid; when Lewis heard what I was up to, he took one whiff and said no. He also said that I might have discovered a new military weapon. Do not distill hot peppers unless you are connected to a gas mask.

Chili Powder

Blend dried, crushed, hot red peppers, such as jalapeño, with cumin, coriander and oregano. Use about six parts hot pepper, two parts cumin and one part each coriander and oregano. Color with turmeric. When combined with stewed tomatoes, sautéed garlic, onion and green pepper, and salt, chili powder is transformed into a zippity-do hot sauce.

Homemade flavoring powders, sans salt, or crushed mixtures of home-grown dried herbs, bottled prettily, make thoughtful gifts for people who are on low-sodium diets.

Cinnamon, *Cinnamomum zeylanicum*, and **Cassia**, *Cinnamomum cassia*, are the oldest recorded spices, yet their bark is as popular today as if it were the newest style. Cinnamon's mission has not changed: a picker-upper for insipid food. In the 1600s some European city dwellers supported professional saucemakers who cooked pre-mixed sauces. A matron who found herself short on flavoring had only to trot down the street with her little beer bucket and pick up a pot of sauce.

Although some commercial distillers extract root and leaf squeezings, the product is less flavorful than bark distillates. Cinnamon bark distills readily in a boiling water still and distilled cinnamon water is handy for use in pickles, drinks and puddings; however, because bark keeps so well, I do not distill cinnamon except for use in fragrances. Cinnamon vapors rise betwen 170° and 212°F (76° and 100°C), and cinnamon oil is soluble in 160 proof alcohol. Actually the spice, cassia, has virtually replaced cinnamon in the United States. The essential oils are similar. The Southeast Asian cassia is a coarser and stronger flavored bark, and both cassia and cinnamon are called cinnamon by the U.S. Food, Drug and Cosmetic Act of 1938.

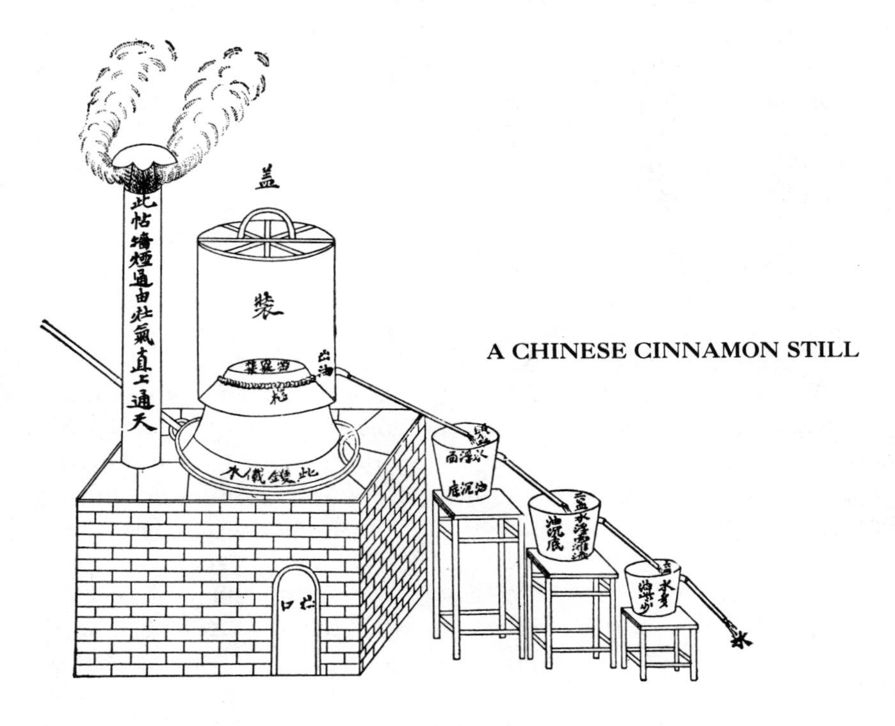

A CHINESE CINNAMON STILL

When the Portuguese sailed around the horn of Africa in the 1400s and discovered a route to the spice islands, they returned to Europe with cinnamon. Later they went back to Ceylon and blackmailed local princes into giving them landing rights. In time they had a stranglehold on the islands and stripped whole areas of wild cinnamon trees by demanding tribute from the natives in bark. The dried inner bark of shoots was cut from the smallish trees, which required deep sand, lots of rain and tropical temperatures in order to grow. By the mid-1600s the Dutch muscled in on the Portuguese and controlled a monopoly on cinnamon until the British East India Company forced them out, and began cultivating new trees. By the mid-1800s more than 35,000 acres were planted in cinnamon; today Ceylon still produces cinnamon, but its near-twin cassia holds a larger portion of the market.

Cloves, *Syzygium aromaticum*, the inflorescence of evergreens indigenous to the Philippines but extensively cultivated in Tanzania and on East African islands, consists of air-dried, unopened flower buds. Primitive cultivation of the tree required that the seeds for planting be transported to the nursery under water and that seedlings be grown for four years in the shade before being transplanted. The tree begins bearing at about eight years, and clove production does not reach its peak until age 25. The flower buds are usually picked in two pickings just before they open. Harvesting is done by hand and the buds are spread to dry on special flooring. When the cloves become dark brown, they are cleaned and packed for sale. Seasoning cloves have a four-sided stem and a calyx with four sepals, but sometimes dried ripe clove fruits are sold as the spice. These are less flavorful.

In 1945 I was assigned to teach at Quigillingnok near the Bering Sea. That primitive village of easy-going fishermen and subterranean homes was accessible only by air. When I was delivered to the flat beach with all my worldly goods about me, I was certain that I had arrived at the end of the earth; however, grinning villagers soon emerged through the high-tide debris, and as the land shuddered with the plane's take-off convulsions, the shore exploded with the scent of cloves.

On the single path through moss to town, which appeared as a mirage of earthen humps and one white building against the horizon, I looked for some plant responsible for the fragrance. None gave off the scent. It was a few days later when the villagers gathered in the school house that I realized that I was surrounded by clove junkies. Everyone chewed cloves. Where did the cloves come from?

I discovered the source when a man, whom the natives called Lanky, an impulsive miner from Good Hope, found me. Entering my one-room school, the wind blowing before him all of his hair that was not tied down, the strange man marched in and handed me a bag of cloves. In his excitement the poor geezer could not talk; nevertheless I read the message. The cloves were a love token. I had never seen the guy before, but it took only one look at his fever-bright eyes for me to start running. It was a small room and I had no where to run. Swinging the clove bag against his advances, spilling them, slipping on them, I finally maneuvered the man into the vestibule, locked the door and began to shake. The smell of crushed cloves drilled into my senses.

Apparently the miner's courage left him as swiftly as it had boiled up. I heard the outer door close. No word had been spoken and I never saw him again. Two reasons: I was transferred and he had his head taken off. He was accidently decapitated by a shovel operator at the platinum mine across the river. To this day I cannot smell cloves without remembering that brief frolic in the schoolroom, and I shall not forget reading that Lanky bequeathed his several hundred thousand dollar interest in the mine to his dog team.

Where did the cloves come from? Nobody knows. Some say that a three-masted ship had washed onto the mud flats near Good Hope and that Lanky had salvaged the spice; one man told me that early day Russians had abandoned the cache when they withdrew from Alaska. The villagers grinned when I asked them, agreeing that yes, they had traded seal skins for cloves, "money you can eat," *auksruk*, they called the spice. *Auksruk*, *clavo*, *gewurznelken* or cloves, this spice has always been popular.

I have not distilled cloves, but the literature indicates that they are rich in volatile oils, yielding from 16% to 19%. Distilled by boiling water (often salted water) or by steam, the constituents in the oil of clove begin to vaporize at about 122°F (50°C). Oil from freshly crushed cloves becomes soluble in 140 proof alcohol, and some modern distilleries are reported to extract clove oil through fractional distillation of the saturated alcohol. Clove distillate or clove oil is used in the blending of brandies, by confectioners, sausage makers, and of course in commercial manufacture of mincemeats, pumpkin pies, buns and spice cakes. Oil of clove is also employed in cosmetics, medicines and perfume.

One of the nicest uses of whole cloves is in brandied peaches. A friend of mine allows a pound of small, peeled but unstoned white peaches and two tablespoons (90 ml) sugar for each fifth of brandy. Poking each peach with an aromatic stud, a whole clove, she places a layer of peaches and a sprinkle of sugar, peaches, sugar, etc., in a canning jar and fills the jar with brandy. She seals the jar and stores it in a cool, dark place for six months. The clove adds a piquant flavor to the brandy, and the peaches are luscious with vanilla ice cream.

Ginger, the *Zingiberaceae* family roots, were said to be popular in south Asia since man cultivated legumes and needed something to spark their taste. Now a healthy supermarket ginger root may be cultivated as a decorative bamboolike house plant, snipping a nip when needed; or the root may be kept by immersing it in a jar of vegetable oil. The oil absorbs the flavor and makes a lovely flavoring in salad dressings. Candied ginger, which may be nibbled or substituted for fresh ginger in most recipes, can be made by slivering the root and mixing it with an equal quantity of sugar. Add a little water to lubricate and boil gently until the moisture crystalizes and the ginger clears. Separate and dry on wax paper. Dust with sugar and store in a jar.

Ginger sauce adds sparkle to low-salt diets. Sauté a handful of chopped onion, a clove or two of minced garlic, about ¼ of a crushed dried hot pepper and a teaspoon or so of minced ginger root. If you use distilled ginger water do not sauté the fresh root. Mix the briefly cooked flavoring into thickened chicken stock, or add to sour cream as a dressing for rice.

Traditionally a product of the East and West Indies and of West Africa, ginger is now being grown in Australia, where producers are improving special varieties.

Years ago I remember Uncle Frank distilling ginger root in his boiling water still. He half-filled his small still with slivered root, covered it with water, attached the head and fastened on the beak, which led to the condenser tube. After lighting a fire under the kettle, he said he could not take his eye off it because ginger "comes over in a blink of an eye." He caught the condensate in a pitcherlike vessel with a spout emerging from close to the bottom and emptying the lower layer of the watery condensate through its mid-high spout. That which stayed in the collector was the more concentated ginger oil. It smelled grand. Although he sold distillates to the vinegar works for pickles, Frank was said to have traded some of his extract to gin mills, where it was added to local hooch. The danger of drinking too much ginger-flavored gin was Jake Leg, a walk reported to be a dead-giveaway that a person was on illicit juice.

If you peel ginger root, do so gingerly. The volatile oils lie just below the skin. Ginger oils start to vaporize at around 167°F (75°C), with the most potent liquid seeming to come over just as water boils. Ginger is not very soluble in alcohol. Surprisingly, my ginger water distillate was not bitey like the root; there must be a gnome with Jake Leg in my still because something happens to the pungency on the way to the collector.

Mustard, *Sinapis alba* and *Brassica nigra*, comes with white and black seeds, plus shades of yellow. Like other members of the *Cruciferae* family, when plowed under as green manure, they restore nitrates to the soil. If you can find a mess of spring mustard greens, they not only are delicious, but they are reportedly good for the memory. The Romans are responsible for the name mustard, which was acquired when they pounded seed into the must of wine to form a sauce. The English first sold mustard seeds by gluing them into a ball with honey. Later an innovative English matron ground seeds into a powder, which only had to be moistened to form a hot sauce. The sugars of white mustard seed are different from those of the black; black seed powders are explosively pungent when moisture is added. The paste is caustic, but if eaten in small quantities, it is pleasantly aromatic and flavorful. White seed powder has no odor and is mild. Thus "double mustard," which contains a mixture of both, is the best to buy for table use. My grandmother bought double mustard flour and used elderberry wine vinegar to prepare her sauce. Turmeric was mixed with the sauce for coloring, and a little salt was added for depth. This is essentially the same as today's prepared mustard. Vinegar helps to preserve the sauce, but mustard itself is said to retard bacterial growth.

Chinese mustard is a mixture of crushed black mustard seed with brown Asian varieties plus a little water. Water is the catalyst, and when moistened, Chinese mustard should be used within the hour or the flavor dissipates.

French mustards are mixtures of white and black mustard seeds that are not completely ground into a powder. Verjuice, an unripe grape juice, vinegar or wine is used as a carrier; and honey, a mince of hot red pepper and tarragon are added for flavoring. Some recipes call for cinnamon, but to our taste the fragrance of the spice confuses signals.

Black and white mustard seeds are usually not distilled for their oils; the oils are fatty, and when commercially extracted by expression, they yield 30% to 40% fixed, or oily, oil. Distilled mustard oil is formed by the fermentation of the ground seed remnants after the fatty oils have been pressed out. The crushed seed cakes are mixed with warm water and fermented at temperatures below 70°F (21°C). As soon as fermentation is complete the liquid is distilled.

A bachelor skier, a Norseman with a wild thatch of tow-colored hair who had a penchant for skiing down upon girl companions and tussling with them in the snow, invited my girlfriend and me to a moose feast at his cabin on Kenai Lake, Alaska. Though house neatness is not my strong suit, this man's organization was cyclonic; we had trouble finding chairs under his debris. As a moose chef, however, his talents were outstanding. First our host showed us a boneless six pound (3 kg) square of meat, next he emptied a quart of prepared mustard into a bowl and rolled the roast in it. Finally he patted on salt, covering the meat heavily. He moved it around gently in the salt with his hands so that it looked like a snow ball. No meat nor mustard showed when he gingerly laid the roast into a flat roaster and baked it in a very hot oven, 500°F (260°C) for about 2½ hours while we cross-country skied on the slope behind his clearing.

Using a towel, our ruddy-faced friend removed the salt-encrusted meat onto a board and cracked off the salt, while we located the table. Swinging the naked meat onto a platter, the bachelor grinned proudly as succulent juices oozed from the first slice. He should have been proud; we all agreed that we had never tasted such an agreeable feast of moose.

Since those days of carefree abandon I have read of beef being similarly roasted in a mustard and salt cast. In fact, just a few years ago a Virginia housewife won a beef contest employing a like process. I have not experimented. If we ever butcher our own ox I intend to try the bold Norseman's technique; if for nothing else, for the memory of tussling in the snow.

Pepper, *Piper nigrum*, the round berry from the evergreen East Indian vine, has worn the crown jewels of spicedom for over 3,000 years. Though challenged by allspice and the capsicums of the chili pepper family—the West African amomum and the Hungarian paprika—black pepper has retained its throne. This monarch of spices has been used as money to pay strumpets, for ransom, rent and as dowry when a man had many daughters and no cash. Black pepper also gave its name to the original black market. Middle East middlemen peppered the palms of hijackers and pirates in an effort to keep pepper flowing across their sands and docks. However, when Portuguese explorers sailed to find the Spice Islands and bumped into the damp jungles of the south Indian coast, they brought economic ruin to the Arabian pepper trade. Lisbon enjoyed being the richest pepper port in all Europe until the Dutch acquired possession of the pepper-producing lands. Thanks to British seamanship, England gained control of pepper, dubbing it corporal of the ordinary table along with salt. Finally, Boston traders took their cut out of the pepper pot, and with the skills they learned in the trade, they founded the Yankee merchant marine.

Both black and white pepper are collected from the same pepper vine. Unripe but fully formed fruit are hand picked and piled to ferment; the heat created destroys the bacteria that can hinder pepper berries from drying. Later, with exposure to the sun, the berries turn into black peppercorns. A few days after the black pepper harvest, the pickers return to gather the yellowish pepper berries just before they turn red ripe. These are soaked in water, pounded to remove husks, and the light-colored inner berries are dried to form white pepper as we know it. White peppercorns are stronger-flavored than black, but the black ones contain a more robust aroma, thus for the best of all pepper worlds, combine the two and grind as needed because their oils oxidize readily.

Pepper oils are commercially recovered by steam distillation, volatilization occurring between 170° and 310°F (76° and 154°C) and are used in canned and frozen foods, in sauces and sausages. I have not distilled pepper. We enjoy its fragrant pungency on French Steak with a sauce made by crushing several spoonsful of black and white peppercorns and mixing them with minced garlic mashed in olive oil. The spiced oil is worked with the fingers into the meat and allowed to permeate the steak overnight or for a half-day in the icebox. The steak is grilled in a very hot iron skillet, turned, seared for no more than five minutes on the second side and served immediately.

Tarragon, *Artemisia dracunculus*, is an odd-fellow of the *Compositae* family that grows in middle Asia. This perennial wormwood cousin loves warm, sunny, dry climes and should be picked in mid-summer for drying or distillation. The tarragon tribe does not seed well, so a few roots should be heaped over with soil to protect them over winter. The herb is a heavy, sweet musky plant said to cure snake bites and demons of the mind. There are two major varieties: French tarragon, a small, deep-green plant with a more delicate flavor, and Russian tarragon, more robust in both taste and appearance. It is easy to become a tarragon addict; like enjoying beer, when you become hooked on tarragon, things go better in its company. With chives and chervil, fresh tarragon leaves lend a whimsical note to omelets. Mild-flavored vegetables are charmed by a dressing of tarragon in sour cream; baked tomatoes, stuffed with tiny, butter-tarragon croutons, are lovely; and tarragon infused in vinegar—some people say there is no more provocative salad dressing when mated with olive oil and a brush of honey.

I have not distilled tarragon, I fear that its gentle heart could not stand the heat. The literature states that volatile oils are recovered in vacuum stills after the herb essence has been dissolved in 180 proof alcohol.

Turmeric, *Curcuma longa*, is a tropical plant that produces the yellow Morroccan powder that is made from the acrid root, which resembles ginger root. When fresh, turmeric is teasingly aromatic and has a musty pepper flavor, which calls to mind the mildew odors of a deserted house. Once ground, turmeric soon loses its zip and should be kept in a dark container because light emasculates its taste. For curries, turmeric adds color and a comfortable depth that blends and gives body to other spices. Turmeric oil is distilled under reduced pressure by steam, but calls for its essence are limited because a major part of turmeric's value is its coloring qualities.

My grandmother sprinkled turmeric in flour for coating chicken before frying it; at picnics I always knew which was Grandma's chicken by the rich golden color, and I would sneak the piece that flew over the fence last. Grandma also colored boiled rice with turmeric; "Beauty is power," she would say, and sure enough, at covered dish socials her rice was the first to go.

Vanilla, *Vanilla planifolia*, has been the most eagerly sought flavoring in the new world since the Spanish overran Aztec nations and confiscated vanilla beans. Obtained from long, pencil-thin pods from perennial orchid vines, vanilla bean pods must be heaped in piles to ferment before being dried and cured for several months in bundles, when a whitish frost, produced by a lactic acid bacteria, will form. Uncured pods are odorless; fermentation is responsible for the enchanting fragrance and flavor. Mexico produces most of the vanilla used in our all-American dessert, ice cream.

Cured and chopped vanilla pods and beans generally are not distilled; instead they are subjected to an alcoholic solvent (about ⅔ water to ⅓ ethyl alcohol) and perked. That is, the watery solvent is circulated through the material many times until the essential constituents of vanilla—aldehydes called vanillin—are extracted. Vanillin is a mixture of aldehydes formed from the enzyme action during curing. Well-cured vanilla beans contain from 1% to 5% vanillin, found in numerous needlelike whitish crystals on the surface of the bean. Vanillin has been synthesized with chemicals found in pine and coal tars, paper pulp and from eugenol, a volatile oil of cloves and coumarin-containing tonka beans. The Food and Drug people are watching the production of vanillin, however, because some synthetics are not only a weak substitute but they may be toxic.

My grandmother kept vanilla pods in a tightly lidded gallon container filled with sugar. She used the sugar in baking and did not add vanilla extract. Her vanilla beans lasted for years; she simply added more sugar when she took some out. She stored the jar in the bottom of the food closet and I can remember getting into it, sitting on the floor, wetting my finger, dipping and savoring. She caught me once and that was the last of that happiness.

Wintergreen, obtained from the sweet birch tree, *Betula lenta*, which grows in wooded regions of the northeastern states, or the wintergreen plant, *Gaultheria procumbens* were formerly distilled indiscriminately to produce oil of wintergreen. Mixing the two oils is now illegal, even though chemically the oils are almost identical. Wintergreen plant leaves are simmered in water before distillation, and birch bark and buds are distilled raw. Fill the boiling water still three-fourths full of slivered sweet birch bark, plus leaf buds if it is early spring, add water to cover, soak overnight, and the following morning tighten the still head and heat over a direct flame. Sweet birch oil volatilizes at about 218°F (103°C), but enough oils come over with steam to provide a bitter distillate. Added to candies or beverages, wintergreen holds the coolness of northern woods. Oil of wintergreen does not dissolve easily in alcohol.

Nuts as Flavoring

Nuts of any lineage have to be the least understood flavoring of modern times. We recognize them as nibbles or in desserts, when in fact nuts constitute a nutritious flavoring for vegetables. Crumbled on buttered garden produce, slivered with sour cream, minced in mayonnaise, or briefly sautéed with onions and garlic and mixed into plain cooked vegetables, nuts—almonds, Brazil nuts, butternuts, cashews, hickory nuts, peanuts, black and English walnuts—add a delightful flavor to food. Add chopped nuts to stovetop dressing, to spaghetti instead of meat, to Chinese foods, to salads. Especially try grapefruit segments lightly scattered with chestnut or walnut slivers. And of course, coconut goes great with cabbage and orange-slice salad.

Nutty Eggplant

Peel, slice and sprinkle a good-size eggplant with salt and allow to drain for an hour. Rinse, drain, chip and stew with a little olive oil in a tightly lidded pan over a low fire. When soft, drain, mash, season with salt and pepper or a spoonful of soy sauce if desired, and mix in a minced clove of garlic, a half-cup of milk, three eggs and a cup of chopped nuts. Turn into a greased casserole, dot with butter or cheese and bake for about 30 minutes at about 400°F (200°C). I sometimes omit the butter and cheese and swizzle olive oil and a few olives over the top.

Pre-mixed Sauces

Crowding the spice shelves with herbs and exotic plant flavorings are a United Nations array of pre-mixed sauces. Sauces seem to be imbued with old-fashioned nationalism: Italian Sauce, Mexican Salsa, French Sauce Allemande, English Worcestershire, Sauce Maltaise . . . I have tried to think of an all-American sauce. Of course catsup identifies with hamburgers, but catsups are universal. I asked some friends. With one accord they said, "steak sauce." The next thing we knew, we were studying steak sauces. Lewis liked that because he insisted that one could not objectively test each sauce experiment without red meat.

Steak Sauce

In a bowl mix ¼ cup (50 ml) vinegar with ¼ cup (50 ml) soy sauce, add one teaspoon (5 ml) hot pepper sauce, and two crushed garlic cloves. Set aside.

Into a second bowl measure ¼ cup (50 ml) vodka and infuse one teaspoon (5 ml), each, freshly crushed whole allspice, broken-up cinnamon bark, plus a few crushed cloves. Cover and set aside. Ground spices may be substituted but they make the sauce gritty.

Caramelize sugar by putting two tablespoons (30 ml) sugar and three tablespoons (45 ml) vinegar into a small, heavy skillet over a moderate heat. Stir vigorously and cook down until thick.

When the caramel is rich brown, stand back and carefully add ¼ cup (50 ml) tomato pulp, sans skin and seeds. I mash a big tomato through a strainer. (My neighbor added wine instead of tomato and that was a good flavor.) In either case, stir like crazy. Lower fire and mix in the soy-vinegar-hot pepper solution after squeezing and removing the garlic. Bring to a boil and remove from the fire when the sugar is fully liquid. The allspice-vodka mixture is strained and stirred in next, plus a spoonful of lime or tamarind juice to contribute tart fragrance. This sauce gains body as it matures, so store in the icebox.

It is fun to experiment. Infuse a few freshly crushed spices and herbs in individual bottles, heat the outdoor grill, gather a few friends, BYOS (Bring Your Own Steak), provide condiments, the caramelized sugar, plus wine and vodka, and let your friends create their own super sauce. Harmony, faith, understanding: All good things meld when saucy good friends get together.

Tabasco Sauce

Another American sauce that is employed to enhance or boost flavor and give character to food is Tabasco Pepper Sauce.

In commercial manufacture, the small *Capsicum frutescens* peppers originally brought from Mexico are picked when they are brilliant red. They are then crushed together with salt before being put into large oak barrels. Five small holes are drilled through the wooden cover of each barrel, and on top of this cover is spread one-half inch of coarsely ground salt; the holes in the wooden cover allow the fermenting gases to escape, and the cover of salt prevents too much air from getting into the barrel. This pepper mash goes through a fermentation process like that used in the making of sauerkraut; the curing mash matures undisturbed for three years, after which time it is inspected by one of the company owners—the McIlhennys. If approved, the mash is pumped into 2,000-gallon wooden vats, where it is mixed for four weeks with strong, distilled vinegar. Finally, the seeds and fiber are strained off before the sauce is finished and ready for bottling.

Soy Sauce

The sauce that made China great is soy sauce. We know that the delicate flavor of lightly cooked food has helped to create the Chinese image of excellence in the kitchen, but the top dog of Chinese culinary achievement is soy sauce. A rich brown liquid with taste reminiscent of meat, a nutritious, salty and satisfying flavoring with the versatility of enhancing monotonous diets, soy sauce is an ancient, Oriental ferment. Though its origin is unknown, scholars say that when ancient man created flavoring from fermented soybean curd, he copied the fermentation techniques used in fermenting fish sauces. Today most soy sauce production is the result of a sophisticated industry, but the ingredients—soybeans, a little rice, coarsely ground, parched wheat, and salt—are the same as when making sauce was a home activity.

In the making of soy sauce, five different cultures—or kojis—employing yeasts, bacteria and mold are used to ferment the soybeans. Mediums of rice and soybeans are used for growing the microbe cultures. After incubation, the cultures are added to the main ingredients of soy sauce: prepared soybeans and wheat. The different inoculations and growing times take a week or so.

Finally the soupy beans are brined for two or three months before being strained, pasteurized and bottled. Twelve and one-half pounds (6.25 kg) of first-quality soy sauce may be recovered from five pounds (2.5 kg) of soybeans. A second-grade sauce is obtained by reprocessing the strainings from the first run. A yield of ten pounds (5 kg) of second-grade soy sauce may be obtained from each original five pounds (2.5 kg) of soybeans. Sometimes caramel is added to darken the color and increase the sauce thickness. Often a third and sometimes a fourth recovery of soy sauce is made. Finally the residue beans are made into cakes and sold for animal feed.

Salt concentration in soy sauce is about 18%; in addition, some starches are changed to sugars, the ferments produce alcohol and lactic acid, acteic acids and monosodium glutamate (a product of soy sauce).

There is no doubt that soy sauce is the leader of sauces. It attracts other flavors and nurtures them into sending forth their fullest essence. Like other leaders, soy sauce has evolved from a long maturation.

Mature leaders must have the ability to enhance existing situations; they must bring out the best in colleagues; they must give satisfactions. Ordinary people and plain food cry out for good leadership. A leader must illuminate truths that transcend the small potatoes and self-interests in the pot. Real leadership overcomes dullness, apathy and violence by giving vision. In foods, flavoring is indeed a leader.

5
CHAPTER
Fragrance

Since childhood I have been intrigued by smells. Two reasons: 1. Our family has been endowed with generous proboscises; if you have something outstanding, you should use it. 2. Uncle Frank believed that man's behavior is greatly influenced by his sense of smell. Odors ruled his life. People used to say that Frank's nose was the reason that he never married "a good Christian woman"; no girl smelled just right to him.

Why do we enjoy some odors? Association? The scent of warm bread in a snug kitchen certainly calls up thoughts of comfort.

Is fragrance associated with love? The flowers and bees, pow-you're-pollinated concept surely encourages the scented pleasures of togetherness.

Writers say that man's sense of smell is mixed with memory and desire. . . .

The young Arab, haunted by the smell
Of her own mountain flowers, as by a spell
The sweet Elcaya, that courteous tree
Which bows to all who seek its canopy,
She sees called up 'round her by those magic scents,
The well, the camels, and her father's tents;
She sighs for the home she left with little pain,
And wishes, even with its sorrows, to be back again.

THOMAS MOORE

Psychologists report that odors play on olfactory glands like notes of music; some delightful melodies of scent trigger yearnings, some smells call up frightful repulsion.

Near Seward, Alaska, walking alone, following a creek in a gentle drizzle, I parted a thicket and came nose to nose with a bear. The awful smell exuding from the place caused me to stagger and I started to run home. Looking behind, I noted that the bushes were unmoving; and with my throat tied in a knot, I inched back. Almost nauseated from the overwhelming stench, I discovered the bruin hanging by his neck in a woodfall trap. He was nearly dead, able only to roll his eyes that bulged above his swollen muzzle. His hind claws, which barely touched the ground, had raked the area clean, his front legs hung limp.

How to get him down? How to keep from fainting? Breathing deeply, taking that foul air into my lungs, fighting light-headedness, I skidded boulders to form a cairn under the animal before pulling down on the bent-over tree top of the gallows. The bear was lowered until he lay in a stinking heap on the mound I had made. I was holding the sapling to the ground uphill, and when at last I saw him move, despite the vile odor, I felt exhilarated.

Abruptly I realized my mistake. How could I loosen the noose from his neck and simultaneously hold down on the tree? Stupid! Should I let go? Foggy queasiness from that repulsive smell curdled my brain until I knew that I could not let go and allow the poor thing to hang. The tree became my grail; some inner sense compelled me to hold the tree low to the ground.

Mists turned to droplets, the drops to rain, but I could not force myself to let go. I sat uphill from the awakening animal holding pressure on the branches. The mud oozed cold, the odor permeated all of my being, the afternoon wore on.

Suddenly the bear got to his feet and to my astonishment I saw the rope dangling on the ground as he half-tumbled into the stream below. For a long while he stood unsteadily, now and then mouthing water. He did not look around and I did not move from holding my treetop in the mud. Gradually, the vile musk smell diminished as the bear staggered across the creek and disappeared into the brush.

How he had freed himself is a mystery, but his odor was no mistake. Even now the hairs on my neck prickle as I remember. I am afraid of that smell.

Though the physiology of smell is not fully understood, authorities agree that the olfactory mechanism, located in the upper part of the nose, consists of hairs or filaments that are connected to cells projecting close to nerve endings. Uncle Frank explained that odors are relayed by the vibrations of gasses that substances emit. The sense of smell is due to the action of these gasses on olfactory relay cells, he said.

In his youth Frank had sailed on a whaling bark in the Arctic; one winter, after following open water leads, his ship was "shut in," held tight in the ice; that cured him of zest for the north. Later he shipped on "spicers" to the Orient. He related magical tales of teeming cities, palaces, temples, curious brown people in fanciful villages and donkey carts piled high with fragrant wood. Naturally he was my childhood hero. A gangling man, big-boned, endowed with somewhat humid eyes and hair that stood on end, Frank was as energetic as a sandpiper; idleness to him held some kind of superstitious terror. He worked as a fireman with my grandfather and also as a distiller of aromatics for the vinegar works. On the side he studied plants and perked off experimental fragrances.

He believed that the abuse of any of man's senses, smell, taste, sight, sound or touch, fed lust. Excesses created a bondage to illusionary delights of life, he repeated. Frank believed that freedom from illusionary cravings could be achieved by attention to two things: respect of all men and of their God and a concern for the health of the earth. "Healthy earth odors rise in tribute to God," he said.

Aromatics of Ancient Jews and Babylonians

Ancient Jews associated aromatic scents with purification. They pulverized the fragrant woods: balsam, cedar and frankincense for incense. With the upward curling smoke, oblations were given in hope that God would be pleased with the scent and hear men pray. Holy oils, made by soaking herbs such as lavender, calamus root and marjoram in olive oil, were given as gifts of thankfulness or as petitions; the presentation of fragrant oils was seen as an outward sign of respect. At one time maidens were anointed with scented oils for several months before they were presented to their grooms.

Babylonians built eight-story altars on which they burned aromatic woods to please their deities; spices and hardened sap of frankincense were thrown on hot coals. Some men of antiquity became so enamored of fragrance they pumiced their bodies so that their skin would accept scents; they also plaited strings of perfumed reeds in their beards.

At feasts, doves that had been dipped in perfumed waters were set free to flutter above the tables to spread fragrance (and no doubt other contributions). One wealthy Persian was said to have perfumed his feet saying, "Why perfume the head and let such delicious scents rise for the birds to enjoy?"

To this day, women throughout the Middle East have been known for their excellence in the personal arts of perfumery. Their customs have been handed down and have influenced the fashion of nations. They believe, as do many women, that scent, the fragrance of Judea, is the key to profound womanhood.

Essential Plant Oils and Their Recovery

Most perfumes consist of three things: essential oils (either natural or synthetic), a carrier to dilute the oil, and a fixative to hold a fragrance and to slow evaporation.

Plant fragrances are due to the minute traces of essential oils that exist in different parts of the plant. Essential oils are found in the flowers of carnation, clove, hyacinth, heliotrope, mimosa, jasmine, orange blossom, rose, violet and ylang-ylang; the leaves and stems of rose geranium and patchouly; the barks of cinnamon and cassia; the woods of cedar and sandalwood; the roots and rhizomes of angelica, sassafras, vetiver, ginger, orris and calamus; the fruits of bergamot, lemon, lime and orange; the seeds of bitter almond, anise, fennel and nutmeg; and the gums of myrrh, olibanum, storax and tolu.

In general, fragrant plant oils, called essential or volatile oils, are present in young plant organs. They accumulate until blossoming time. With the appearance of flowers, the alcohols and esters of volatile oils are converted to oxygenated products, the aldehydes or ketones. The flower becomes more fragrant as it readies itself for fertilization. Some botanists have declared that, once concentrated for the part they play in fertilization (during pollination some oils are consumed), the essential plant oils take no further role in the plant's life processes, but other studies show that essential oils continue to assist with the plant's metabolism during the seed formation period. To be on the safe side, most aromatic plant parts should be harvested just before fertilization.

Essential oils may be light and volatile as alcohol; a few are waxy or heavy; and most leave a mark on paper that soon evaporates. Essential oils may be colorless to dark brown, liquid or solid and all are inflammable.

Perfumer-chemists understand that very few plant oils have a pleasant odor in the concentrated state. Some essential oils are said to be toxic. Some are narcotic and numb the nose. Only after the oil has been diluted does the odor become agreeable. Many plant oils contain numerous components, and those with similar chemical makeup have similar smells. Chemicals in a plant oil often volatilize under different conditions, and some components form a mix called an azeotrope. The azeotrope sometimes volatilizes more quickly than the oils of the single components in the plant oil.

The volatility of essential plant oils varies. The true boiling point is generally very high: 320° to 550°F (150° to 288°C), yet strangely, many plant oils volatilize below or with boiling water. Though essential oils are only slightly soluble in cold water, most oils dissolve in boiling water in sufficient amounts to impart their odor and render the decoction fragrant.

A young lady asked me, "Do you have to use a still, or can you make fragrances on the top of your stove?"

A FRENCH PERFUME FACTORY

If you can catch the vapors from boiling orange blossoms or roses when you heat them in a pan on the top of a stove, you could make your own watery fragrance without a still, I told her. However, if you wish to capture a concentrated product, you must catch the vapors and separate them from most of the boiling water. And because distillation is a separating process, it is a much better method of trapping fragrances. We went on to talk about extraction of plant fragrances, and I told about plant oils dissolving in alcohol. She asked if scents could be captured in a solvent, that is, could a person soak an aromatic plant part in alcohol and make a fragrance? It is possible for fragrance to be captured in solvents, I replied. However, numerous plant soakings are needed to saturate a solvent with scent. During the recharging of the solvent by repeated macerations (soakings), there is scent loss. Volatile oils evaporate in a wink, I told her, and some change odor when exposed to air.

A CLASSICAL STILL WITH A WATER COOLED HELMET

Traditionally there are five ways to recover essential oils from plant substances: 1. expression; 2. extraction by means of cold enfleurage in solid fats or by maceration in warm fats; 3. immersion of plant materials in solvents (for extraction in commercial stills); 4. boiling water and steam distillation; 5. a combination of these processes.

Expression. Squeezing the essence, as in pressing oils from a lemon rind, is called expression. In perfumery, bergamot oil is expressed from citrus rind, *Citrus bergamia*, and is extensively used in eau de cologne.

Extraction. An ancient form of scent extraction, called enfleurage, was accomplished by repeated exposures of flowers to fats, until the fats had absorbed fragrance. The fragrant fats were called pomades, and in olden times they were used as hair and body cosmetics.

In more recent years the flowers that were to have their essence extracted were tied in a mesh cloth and immersed in purified, molten lard-beef fat or in olive oil. The fats were kept between 122° and 158°F (50° and 70°C) and the time of immersion was about three or four days. When time-in-fat was up, the mesh bag was squeezed, the spent flowers thrown out and the same fat recharged with fresh flowers. That was repeated about ten times. The fragrance was then extracted from the warm fat by shaking it with an equal amount of 180 proof ethyl alcohol, refrigerating and skimming. The fats were used in soapmaking, and the floral essence remaining in the alcohol was employed as a handkerchief perfume or added to other fragrances. If olive oil was used as the floral absorbent, the fragrant oil, called Huiles Antiques, was used straight as a pomade or integrated into skin creams.

Enfleurage or maceration in warm fats does not seem to work with the use of modern hydrogenated fats or with polyunsaturated oils; I tried several different kinds, and zilch, no scent had been absorbed.

Even the slightest heat affects the aromatic properties of most flowers, but some blossoms (jasmine, tuberose, jonquil and lily of the valley) can stand no heat, thus they are not distilled. These flowers contain very little ready-made essential oils; instead they produce perfume as a part of their life process. If these cut flowers are placed over fat, they continue to live for a long time; they emanate odors that are absorbed by fats at room temperature. Their fragrance is extracted from the fat by alcohol.

Solvents. A commercial means of extracting essential oils is the immersion of plant parts in volatile solvents, such as alcohol or ether. This method is not practical in an ordinary home. The most commonly utilized commercial solvent is diethyl ether. The flowers that are to have their volatile oils extracted by means of solvents are lowered in mesh baskets into airtight vessels and sealed. Ether solvent is pumped into the vessel, where it remains in contact with the flowers for six to eight hours. The flowers are replaced by fresh ones several times until the diethyl ether is saturated with the essential oils. The ether is then distilled in a vacuum still under reduced pressure (and consequently at a lower temperature that does not spoil the floral odors). This process so concentrates essential oils that they are caustic and must be diluted before sale.

Distillation. Boiling water and steam distillation are the most popular methods of removing essential oils from plants.

The plant material to be distilled by the boiling water method is put on a rack in the copper boiler, water is added to cover the material, the head of the still is secured, the condensing tube is connected to the head, and a check is made to see that cold water is running around the condensing tube. The still is fired and steam leaks are clamped. A narrow-mouthed collector is placed under the condenser exit.

As the water boils and begins to force vapors into the head of the still, a humming sound vibrates the metal. The pitch mounts as steam pressure builds, and when the first vapors work their way through the copper tubing into the condenser, crackling and snapping noises alert the distiller that vapors are reluctantly contracting as they meet the cold surfaces of the condensing coil. Very soon a spurt of foam and liquid will explode from the end of the condenser. After the "first shot" is fired, as the first pipe-cleaning spurt is called, a collector is placed under the condenser exit and liquid will perk from the tube with a regular cadence.

A distiller rarely distills more than one-quarter of the amount of water that was put into the boiler, usually much less. He smells the distillate, fingers it, and when it begins to run watery both in aroma and feel, he shuts down the still. Very often he recharges the boiler water with fresh plants. The process of recharging the same water with new material is called cohabation, and it may be done as many as six times. The good stuff that is in the collector at the end of the condensing tube may be returned to the still on each recharge of fresh plants. Though I sometimes cohabate material, I do not return my distillate to the boiler. Tightly capping the distillate when the run is done is important.

Boiling water pot stills have been utilized in most countries for the extraction of numerous essential oils, including roses, rose geranium, lavender, mints, camphor, sandalwood and fragrant grasses. Boiling water stills are practical in remote areas where transportation of flowers to commercial stills would take too long and harm the product.

A 19TH CENTURY STEAM STILL FOR AROMATIC PLANTS

Steam distillation undergoes essentially the same process as boiling water distillation, the difference being that steam is produced in a separate boiler away from the still. A small amount of water is poured into the still proper, a smaller vessel than the water boiler, and the raw material placed rather loosely over the water to fill the vessel three-fourths full. Steam is piped into the still through the bottom in metal commercial stills. In small glass, laboratory-type stills, it enters the still through a tube inserted into the stopper, and the steam tube extends to near the bottom of the still. Thus the steam emerges from below, perks up through the plant material and volatilizes the plant oils. The steam forces vapors to exit through a short second tube located in the top of the still, and they are directed to the condenser. The condenser may be a coil immersed in cold water or it may be a one-piece water jacket containing cold running water that encases the vapor condensing tube. Vapors crackle and pop as they are cooled in the condenser, but finally they give up their vaporhood and contract into droplets of liquid that run out of the bottom of the condenser tube into a collector.

The home distiller must constantly test his distillate by feel and smell because many plants are quick to capitulate. Their volatile oil sacs rupture as soon as steam heats them; they give up their oils that mix with steam and are often carried into the condenser before or with boiling water. To test distillates: When the feel loses its slickness and the odor changes or becomes less strong, it is generally time to close off the steam, remove the still from the heat and cap the collector.

Steam distillation is practical when working with small quantities of plant material; a steam still is also useful in redistilling distillates. When redistilling a liquid to concentrate it, steam is sometimes employed, but often the steam tube is disconnected and plugged and the liquid is distilled over a low fire. Second-time distillation is even faster than first extraction of plant oils. When redistilling a fragrance, a perfumer generally recovers only the first one-quarter of the liquid, which contains most of the essential oils. That which remains in the still is mixed with the residue from the first distillation. These watery distillates are settled, filtered and employed as an aromatic water in formulas calling for distilled water. The stronger distillate and the aromatic water are stored in completely filled dark glass bottles with foil-lined screw caps. Corks may be used, but rubber stoppers or plastic lids may absorb the scents of some substances.

Steam stills generally have two flasks, but I saw an old, custom-made three-flask steam still in Hope, Alaska, in 1951. It was mid-summer, and my friend Suzanne and I had taken her new car for a drive north of Seward because we wished to explore the abandoned mines in the mountains above Hope and Sunrise. We had been told that gold had been found in the area before the Klondike rush and that 15,000 hopeful miners once had swarmed the precipitous slopes.

As we drove into the mountains, the gravel road became dirt and the dirt became a brush lane and there was nothing between us and the rounded peaks softly silhouetted against the throbbing indigo sky. Here and there among the rocks that circled our horizon we could see black holes that were mine shafts, the remnants of 50-year-old shattered dreams.

When the lane abruptly ended at a lake, Suzanne and I shouldered packs and began to hike. Nowhere on this wondrous planet exists the languishing, penetrating serenity of an uninhabited alp. As we picked our way higher toward the black mine shaft openings, we paused to look down on the green bowl-shaped highland with its incredibly blue pocket lakes. Suzanne's car was a toy; to the north, the sea arm, Turnagain, was a glistening steel knife.

Poking into the deserted mines was a visit to the dead. Poorly defined shadows reached toward us, the tunnels appeared to be illuminated with yellowness, water dripped and hoarsely echoed from every quadrant, globs of mold clung to tortured palings, and around us and everywhere icy drafts swirled up from the mountain's bowels.

It was not long before Suzanne and I had our fill of the spooky place. We retreated toward the car, now glowing brightly in the sunshine about three miles away.

Suddenly, from a clump of alder protruding from a ravine, a man's voice rang out. It seemed to be so permeated with agony that we both stopped in our tracks. We found a fisherman, dead drunk and lying in a tent beside a stream. He was a corpulent, untidy individual in his middle years who had a week's beard, a bald brow, large slack lips and was clothed in filthy dungarees.

Whether from pain or inebriation (empty bottles littered his camp), the man muttered insensible words and held out a horribly infected thumb. The shank of a fishing hook protruded through the thumbnail.

What to do? We could not get him to stand, we could not carry him, we would not leave the man unattended; so Suzanne and I decided to remove the hook, make him as comfortable as possible and get help. First we scrubbed the grimy paw (the man moaned but seemed unaware of our presence). Next we sterilized a razor blade and with one swipe slit the restricting part of the thumb nail, and in the gush removed the fish hook.

Washing the wound a second time with alcohol (his gin) we bound his hand, left food nearby, covered the man and raced for the car.

Once back at sea level we turned toward Hope, parked in front of the general store and alerted the populace to the emergency. Townsmen, who said they knew the man we described, jumped into a truck and rattled off toward the high valley to help him.

Suzanne and I were alone with the storekeeper, an aged man who walked with the same tilt to his frame as that of his rickety store building. None of the six or seven standing buildings in town stood plumb with the world; they all had been heaved by frost or made lopsided by quakes.

When we were getting into Suzanne's car to drive home, the storekeeper volunteered that he had a new car, too. He nodded toward a tipsy building across the street and offered to show the car to us. It was apparent that the old timer was in love with his vehicle, and not wishing to diminish his pride, we followed him. Never had a 1928 coupe looked more grand to the eyes of a man. "Just see what an elegant bird I've captured," his face seemed to say as he showed off his prize.

While Suzanne inspected the automobile, I wandered to the back of the storehouse. Gracing one shelf was a three-part glass steam still that resembled three coffee percolators with long beaks. The largest had a one-piece glass head with a spout that reached diagonally downward to enter a second glass flask near its base. The second flask apparently was the still and the first was the water boiler that produced steam. The second flask head (which was clamped in place with metal snaps like those that held the larger boiler head in place) sported a longer and narrower beak, which passed through the third short, fat glass coffee pot. Rubber gaskets held the extended still-beak in place as it ran into, and passed through, the third flask.

Answering my inquiry, the storekeeper said he used to fill the third flask with glacier ice; it was his condenser. Steam from the first pot heated the liquid in the second still pot, and the vapors that rose were forced through the glass beak that went through the ice-cooled condenser. The storekeeper showed us where he put his collector under the end of the condenser exit tube. He had had the still made in St. Louis and he was willing to talk on and on about his "new invention," of which he was as proud as his "new car."

The condenser was unique in that it had a glass spout that was lower than the top of the pot; thus melted ice water ran out of the spout and could be directed away from the distilling area.

Suzanne and I wished to question him more about the steam still, but we had a long drive home and could not tarry to talk with the man who had come north with the gold rush. We always meant to return and visit the storekeeper of Hope.

The advantages of steam distillation are that material is away from the direct heat, the possibility of scorching is eliminated, the whole unit is smaller, and the process is quicker because the distiller can distill one substance after another using the same boiling water. The major drawback with a glass steam, laboratory-type still is that large volumes of material cannot be run. One time I sketched an idea for a big steam still; Lewis took one look and said that if I move that Rube Goldberg into his workroom,—a shed—he was going to drop my favorite cat into the first batch I tried to run off.

Chemistry of Fragrance

Chemistry plays the major part in odor. Some smells are flowery (heliotrope, rose or vanilla), spicy (spice bush, cinnamon), fruity (orange), and some are resinous (pine needles). Different smells are the result of the chemical makeup of the plant oils. Some chemicals give off organic, flowery fragrances that are generally carbon compounds, most containing hydrogen and oxygen. The alcohol constituents of plant fragrances (those ending in -ol), such as geraniol found in roses, contain carbon, hydrogen and oxygen. The aldehydes of plant oils (those ending in -al), such as citral found in lemongrass, contain partly oxidated alcohols. The ketones (ending in -one), such as irone found in orrisroot, are secondary alcohol oxidations, and the esters of flowers are the alcohols mixed with organic acids, such as the fragrance of lavender.

The terpenes that contain carbon, though generally not aromatic, are components of plants (ending in -ene), such as di-limonene found in expressed oil from lemon peeling. Terpenes are usually found in conifers. Carbon compounds, that contain sulphur, most often have offensive thiol odors, such as is found in garlic.

The chemical analysis of aromatic plant properties enables chemists to synthetically produce fragrances. Chemist-perfumers make fragrances by distilling plant material and redistilling their essential oils. Thus they separate the different components of an essential plant oil and they are able to reproduce them.

Carriers of Perfumes

During a recent holiday I spent a number of days examining modern fragrances. Nearly all popular-priced scents were composed of synthetic substances, and all contained alcohol. Over succeeding weeks I tested fragrances and smelled different every night, until Lewis said he would wake in the middle of the night and wonder whom he was in bed with. My unscientific tests showed that scents using alcohol as a carrier lasted from one to four hours.

To oversimplify: Aromatic plants yield tiny quantities of strongly scented essential oils. These volatile substances, or their chemical duplicates, are the basis of most fragrances. They must be mixed with something to carry the scent and to dilute their strength, generally distilled water or ethyl alcohol.

Alcohol, alternately called rectified spirits of wine, grain alcohol, ethyl alcohol or purified cologne spirit, is of prime importance both in the creation of chemical perfume synthetics and in the manufacture of natural fragrances. Ethyl alcohol, 90% by volume (which would be 180 proof) is preferred by professional perfumers. It is colorless, burns blue, is nearly odorless and is capable of being mixed with acids and water. As a solvent for extracting essential plant oils and in the dilution of fine perfumes, 180 proof ethyl alcohol is recommended.

Ethyl alcohol, which may be purchased from druggists or in most liquor stores as grain alcohol, is manufactured by fermenting organic substances rich in sugars or starches and separating the alcohol by distillation. Even after several redistillations, called rectifications, ordinary ethyl alcohol contains about 10% water. This water in ethyl alcohol does not harm aromatics. Fusel oils, on the other hand, do pollute alcohol; they make it smell like old potatoes. Fusel oils are present in several kinds of ethyl alcohols that have been fermented from organic substances. In order to eliminate fusel oil, the alcohol may be diluted with water, which throws the oils out of solution, filtered through specially prepared charcoal and redistilled before use. An easier, though less exact, way to rid alcohol of fusel oil is by mixing the alcohol with a little commercial vegetable oil, shaking it, refrigerating it for a day and skimming off the oil.

Alcoholic beverage taxes are placed on ethyl alcohol even though it is used in perfumery. Industrial alcohols or metholated spirits are not taxed so heavily, thus you may be tempted to use methyl alcohol in fragrances; do not do so. Industrial alcohols contain 90% crude alcohol made from grain and/or wood, plus 10% wood naphtha and crude pyridine, a poisonous substance with an unpleasant odor.

Isopropyl alcohol is sometimes used in less expensive scents. It is a good solvent but lacks the pleasant winelike qualities of ethyl alcohol; in addition isopropyl may be a little toxic to the skin when mixed with some chemicals. Isopropyl is a by-product of petroleum; when pure, it is colorless, mixes well with essential oils but has a characteristic alcohol odor that may be covered by the addition of commercial terpeneless citrus oils. When not pure, a water-isopropyl mix added to some natural essential oils creates a cloudy product because of the water in the petroleum-base alcohol. The addition of a little terpeneless oil (those plant oils from which terpenes have been removed by distillation in vacuo) often clears the perfume. Thus if you use isopropyl, the addition of a few drops of terpeneless citrus oil, available through your druggist, will cover the alcohol odor and clear the product.

Plain distilled water or a homemade fragrant water distillate may be used to dilute some essential plant oils. As with the utilization of isopropyl alcohol in fragrances, a few drops of terpeneless citrus oil is necessary to deter clouding.

A substance that was once used to extract plant fragrances or to dilute plant oils, but is rarely used today, is acetic acid. Acetic acid is formed when a fermenting brew is allowed to go into a second fermenting, called a bacterial phase, and the liquid is turned into vinegar. The vinegar of commerce is about 5% acetic acid, but 10% or 15% acetic acid is used in perfumery as a solvent for some essential oils. Acetic acid mixes well with ethyl alcohol to produce small quantities of ethyl acetate, a pleasant fragrance.

Glycerine, obtained from fats as a by-product of commercial soapmaking, is another substance used in perfumery. As lye is added to fats in soapmaking, it breaks down the fatty acids and separates an alcohol called glycerol from them. Glycerine is useful in cosmetic perfumery because it mixes with other alcohol, with water, and it sinks into the skin. One caution: Glycerine must be diluted, by no less than its own volume, with distilled water because it is very drying. Aromatized alcohols and/or fragrant waters are often mixed with glycerine, given a fancy name (Oil of Goat's Baa) and sold as a high-priced skin softener.

Fixatives of Fragrances

The third essential ingredient to most perfumes is a fixative. Fixatives are substances with high boiling points. Adding a small amount of a fixative to a volatile liquid in which it is soluble results in the decrease of vapor pressure of the volatile liquid; this slows evaporation. By slowing evaporation, a fragrance does not lose its odor so quickly, thus the life of the perfume is prolonged. Fixatives also tend to bind different odors together and give body to fragrance. Skill as a perfumer comes from blending volatile substances with fixatives that will hold a fragile odor but will not dominate the fragrance. The fixative must equalize the rate of evaporation between various constituents and blend the major scent.

Perfumers classify fixatives as pleasantly aromatic, such as benzoin, or disagreeable, such as civet. Plus there are neutral fixatives. Fixatives or their synthetic counterparts may be purchased from perfume suppliers or from pharmacists. They are generally cut with a solvent when purchased. Ingredients and suggested use are also included on the label.

Traditional animal perfume fixatives were ambergris, a waxy substance found floating in tiny to one-hundred pound pieces along the shores of tropical countries and believed to originate from a pathology in sperm whales. Civet is an excretion from the African civet cat that is more malodorous than skunk and obtained by teasing the cat and catching his squirted defense weapon. Musk is secreted from the male musk deer of central Asian highlands and obtained by killing the poor deer for his pods. The musk loses its heavy odor when dried, but fresh it has been said to knock people senseless. Stories are told of Asians addicted to musk; they killed merchants for it and put beads under their toenails in order to have a constant supply of the odor. Rarely seen naturally, these three ancient perfume fixatives become incredibly sweet when diluted with solvents. All have been created synthetically with chemicals and petroleum products.

Other animals, such as beaver, muskrat, musk-ox and skunk, secrete musk; but their musk has not been commercially promoted as perfume fixatives. I can attest to the strength of oxen musk. In the 1940s I went down the Yukon River in the most odoriferous stern wheeler in the Arctic.

Several years before World War II, our federal officials decided to introduce musk-oxen into our northern climes, hoping the oxen would provide industry and food for the Eskimo. The animals were shipped by freighter to Seward, loaded on a train, transferred to the river boat in Nenana and taken down river toward the Bering Sea lands where they were to be off-loaded and herded by native peoples. The plan backfired. An amorous bull "got-to-showing-off" in the hold of the wood-burner as it steamed toward St. Michael. The animal broke into the engine room, the engineer evacuated the area via a window, got stuck and screamed bloody murder until hands who were firing the boiler with small trees came running to the rescue. When the two Indian boiler-tenders came face to face with a musk-ox positioned between the pistons and their boss's posterior, one man attracted the ox's attention with a poker while the other scooted around the moving mechanism and pulled a steam valve. With the whistle a cloud obscured the action, but the word is that the ox gave forth with a spray of musk that, mixed with the steam, flattened all three men. It also permeated the wood for all eternity. The Captain at the wheel, high atop the vessel, suspected something was amiss when he saw steam gush from the bowels of his flat-bottom craft. Putting his face to the engine room pipe, he got a whiff and hit the deck. The fumes were so overpowering that deckhands from the barges being pushed ahead downstream had to leap aboard and drag the Captain, mate, engineer, purser (who remained intoxicated for a week) and the two hands onto the barges and cut the sternwheeler free. A few hours later the still-prostrate men saw their own river boat pass them with the musk-ox roaming the lower deck. It sailed all the way downriver, the story goes, to Paimute, where it snubbed a bank. The oxen went ashore and were soon eaten by the wolves. The Captain told me this story; I smelled the musk and believed every word. The sternwheeler, now a museum piece, still reeks of that wayward ox.

Some perfumers employ plant fixatives and believe clary sage, *Salvia sclarea*, an easily cultivated garden plant, to be the finest fixative for perfumes. As with other plants, the maturity and handling of the upper herb and flowers, the climate and cultivation varies; thus the product varies. The clary sage aroma, which has been likened to muscatel grapes, lavender, ambergris and bergamot, sweetens and mellows as it ages. Known as "oil of amber," clary sage oil adds holding power to most fragrances. Though it yields very little oil, about 1%, its volatile properties lend themselves to distilling with water. Using a boiling water pot still, immersing the sage and firing gently, vaporization begins early and a strong fragrance comes over between 200° and 208°F (93° and 97°C). Clary sage oil is soluble in 160 proof ethyl alcohol.

Other plants that contain fixatives and are used in perfumes are basil, cassia, clove, cypress, marjoram, orrisroot, patchouly, sandalwood, sassafras, spikenard, thyme, vetiver and ylang-ylang. Most plant fixatives may be extracted by aqueous distillation or by maceration for two days in 160 to 180 proof ethyl alcohol.

Different Types of Fragrances

Over the years I have been perplexed about meanings of words employed in perfumery. What is an essence, or a bouquet? In an effort to clarify my own thinking, I searched commercial perfumer literature for examples and usage of professional nomenclature. Unfortunately, there seems to be a variety of temporal, regional and stylistic labels applied to formulas and nomenclature. The following word usage, however, seems to be most universal among professional perfumers.

An essence, extract or tincture is a fragrance that contains a single aromatic plant oil plus alcohol.

A bouquet, perfume or perfume bouquet is a scent that contains two or more different essences, extracts or tinctures, plus additional alcohol. These compounds are usually diluted a second time with alcohol before being matured in a cool, dark place for a few weeks.

A cologne, traditionally speaking, is much like a bouquet or perfume, except it contains small quantities of a variety of essential oils dissolved in alcohol, all of which are distilled together in vacuo. This premixing and distillation is said to give a unified character to the scent. Cologne is diluted with a higher ratio of alcohol than perfume.

A Guide to Aromatic Plants

Before starting to make your very own fragrances, play around with samples. Start distilling water. Next, prowl the countryside for blossoms, ask permission, then pluck a handful or two. Try pine needles and mint. Heat your still, gradually increase temperatures until you hear the crackling of vapor being formed. When the condensate begins to come over, feel and smell it. Note changes in odor. When the run loses its overpowering odor, shut down and lid your distillate.

The world of scent is infinite; mix, match, cohabate, concentrate, add a bit of commercial oils or fixatives, but always store in completely filled, dark bottles with tight lids and hide them away in a cool place. Decant your best creations before Christmas, tie a bow on them and give them to your friends. Fragrances are gifts that hold the sunshine of summer and the enchantment of a moon-bathed night.

Fragrance can be a riddle. Gill-over-the-ground, *Glechoma hederacea*, smells offensive when you run over it with the lawnmower; distilled and diluted with cedar, sassafras and a bit of orrisroot oil, however, it gives off a refreshing scent. On the other hand, honeysuckle, *Lonicera japonica*, emits an exquisitely heady odor in its natural state, yet if it is distilled by itself with water, its fragrance becomes muddy.

Most of the aromatic plants listed below have had their essential oils duplicated as synthetic fragrances, and most may be purchased in small quantities through drug stores or from cosmetic suppliers. A few wild and garden plants that traditionally have been distilled for fragrance are included together with facts and fancy about old-timey herbs. Like any natural process, distillation varies, every batch is different; variety is what makes home-distilled products intriguing.

Allspice, *Pimenta dioica*, although generally thought of as a flavoring, allspice is also a fragrance that offers a spicy note to cut the clinging sweetness of heavy scents. A little goes a long way. Pimento water may be distilled by the boiling water method, using one part bruised fruit to two parts water. Collect about one-half of the original liquid. Vaporizing just before water boils, 203°F (95°C), this strong-smelling condensate should be included with perfumery because of its versatility and use in men's colognes. In addition to allspice, other flavoring spices used in fragrances include anise, bay leaf, caraway, cinnamon, clove, dill, fennel, mace, marjoram, mint, rosemary, tarragon and wintergreen.

Balm, *Melissa officinalis*, is the sweet lemon balm that grows in herb gardens throughout America. The whole plant may be distilled by laying it loosely on a rack in a copper pot still, covering it with water, sealing the head and bringing the water to a gentle boil. Collect about one-quarter of the original liquid. You may wish to cohabate your material several times to create a very fine melissa water, which is often substituted for citronella in alcohol-based jasmine, lilac, rose and other sweet scents. Melissa distillates provide a pleasantly fresh lemony fragrance used in hair washes and cosmetics. Melissa balm starts vaporizing at about 185°F (85°C); its oils are soluble in 140 proof ethyl alcohol.

Benzoin, *Styrax benzoin*, is not readily available in the United States. This resin is obtained by cutting the bark of the East Indian tree and collecting the secreted tears that form into hard lumps. Benzoin has excellent fixative qualities and a pleasant, vanillalike odor that makes it, or its chemical counterpart, a popular additive in jonquil, hyacinth and orchid artificial fragrances. Benzoin and benzoic acid keep cosmetic fats from becoming rancid. Even the tiniest amount, 1/1000 part by weight, will keep cosmetics fresh smelling. Tincture of benzoin, made by mixing one part benzoin to eight parts 180 proof ethyl alcohol that has been warmed to body heat, is often added to fragrances to give depth. Benzoin is commercially recovered by distillation in vacuo.

Bergamot, *Citrus aurantium*, is a Mediterranean tree that bears greenish, bitter, lemon-type fruit, the thin rind of which contains a fragrant oil that is expressed rather than distilled. Bergamot oil combines well with most floral odors. Natural or synthetic, it is useful in perfumery to give rich body to thin scents. Tincture of bergamot is made by mixing one part oil of bergamot with eight parts 180 proof ethyl alcohol. The fragrance of bergamot is especially nice with lavender perfumes.

Calamus, sweet flag, *Acorus calamus*, is found growing on the banks of ponds in temperate climates, and its rhizomes are known for their peppery, camphorlike odor. In fall or early spring, young, firm roots are gathered, washed, rootlets removed, and the roots skinned lengthwise because most oils are found near the root skin. Root skin slivers are distilled by boiling water or steam, and the soft, mellow fragrance is useful where a peppery touch is needed. Fresh root peelings begin vaporization of essential oils at about 158°F (70°C), and vapors continue to come over with the steam of boiling water. In addition to calamus water, calamus essence may be made by infusing the root peeling in 180 proof ethyl alcohol, one part root to four parts alcohol, for two weeks in a dark, tightly covered vessel. When strained, the aromatized alcohol is helpful in compounds that have a very persistent odor, such as those containing oak moss or sassafras; calamus has a softening effect.

Cherry, *Prunus serotina*, wild black cherry root bark may be peeled, washed, shredded and immediately distilled by steam. Cherry oils deteriorate if allowed to dry before distillation. The fragrance is reminiscent of bitter almond, and distilled cherry water adds a deep, rich note when combined with citrus scents such as orange blossom fragrances.

Coumarin, as referred to in perfumes, is a well-known, new-mown hay scent extracted from tonka beans, woodruff and fellows of the bedstraw family. Apparently coumarin does not always exist in the plants; rather it is formed by fermentation because its odor becomes noticeable only after drying. It is synthetically prepared, largely from salicylic aldehyde, and is soluble in almost all liquids used in perfumery. The extremely light- and heat-sensitive natural oil vaporizes in steam distillation at about 208°F (97°C), and it is soluble in 180 proof ethyl alcohol. Whether the synthetic or natural coumarin is used, this fragrance is excellent when blended with vanillin and heliotropin, and it is often employed with verbena and lemon grass fragrances.

Elder Bloom, *Sambucus canadensis*, *S. niger*, is the flower of an American and European bush that may be distilled in a boiling water pot still charged with salt water. Boil the still vigorously and collect one-fourth of the original liquid. Elder blossom water should be made from fresh flowers because the odor deteriorates on drying. The aqueous distillate cools to a yellowish color, probably caused by the oil, which readily becomes viscid. Though scarcely pleasant smelling when first distilled, elder blossom water acquires a refreshing bloom when allowed to mature. An early-day elder flower fragrance was made by mixing the distilled water with coumarin, rose, palmarosa oil and musk essence.

Frankincense, from the tree *Boswellia carteri* and other species, is called olibanum resin by perfumers. The gum-resin is obtained by making an incision in the bark of the East African and Arabian boswellia tree and collecting the dried tears. The aromatic frankincense wood was originally burned for its fragrance; its residues were pulverized to form the black eye shadow, called kohl, employed by ancient Egyptian beauties. Frankincense was also used as a sachet among clothing and in the embalming of the dead, a practice that the Jews borrowed from the Egyptians. In more recent history, boswellia wood and its resin have been used as an incense.

Incense

Pulverize or grind five ounces by weight (150 g) olibanum, two ounces (60 g) benzoin, a half-ounce (15 g) storax and a half-ounce (15 g) dried boswellia wood. Mix and store in a well-sealed container. To start incense smoldering, light a fire with a bit of twisted paper; if the incense flames, douse it with a wine-soaked rag. This formula was told to me by a blue-eyed, black-bearded Russian Orthodox priest who made his own incense and burned it with great pride.

Ginger, wild, *Asarum canadense*, is dug in late fall, washed, chopped and macerated in equal amounts with common vinegar for two weeks. After filtering, the ginger-vinegar extraction is mixed, again half and half, with 140 proof ethyl alcohol, matured for two months before combining with overly sweet scents such as imitation violet.

Heliotrope, *Heliotropium arborescens*, is a fragrant little plant that endlessly follows the sun. In mythology the tattle Clytie was cast off by Apollo, and she pined, naked and unkempt, on a hillside for nine days. Fed by nothing but her tears of remorse, she gazed endlessly at the face of her lover, the Sun, until her limbs grew into the soil and she was changed into a flower. Early heliotrope perfumes were made by floral maceration in warm fats, and the scent was said to combine the fragrances of vanillin and coumarin. Today the almond-scented synthetic acetophenone reproduces the tenacious odor, which blends nicely with rose, jasmine and tuberose, using benzoin as the fixative.

Lavender, *Lavandula angustifolia* an ancient member of the *Labiatae* family, may be planted from seed or from cuttings in April in temperate climes. The colorful flowers terminate in blunt spikes, with "full picking" obtained from the second to the fifth year. Spikes are cut on a clear, dry day around August first when blooms are fully developed. The flowers are stripped and together with the upper leaves put loosely into a net bag, which is suspended above the water in a copper pot still; or they are layered on a rack in a steam still. The water should be adequate so that it will not boil dry, but the distillate is said to be best if the flowers are not immersed in water. The still is sealed and boiled so that the steam will rise quickly and carry this explosive oil with it. Volatile oils begin vaporizing between 200° and 210°F (93° and 98°C). The still is closed down as the distillate droplets become less fragrant. Fresh or dried lavender flowers may be employed in distillation, but the key is to distill as rapidly as possible. Settle and mature lavender water in a well-filled, tightly stoppered bottle and store in a dark, cool place for two months. Between 10% and 40% lavender oil may be recovered by redistillation and the odor is said to improve when matured for several years after 1%, 200 proof alcohol has been added. Lavender oil is soluble in 140 proof ethyl alcohol.

Fine cologne may be made by mixing half triple run lavender water with half 180 proof ethyl alcohol and adding a few drops of musk ambrette (available through perfume suppliers) as a fixative. Triple run means to rectify or redistill the distillate two times.

Lemongrass, *Cymbopogen citratus*, is a popular plant gathered by East Indian farmers. Lemongrass is distilled by tying it in bundles shortly after the beginning of the rainy season and subjecting the whole bundle to boiling water distillation. After filling large stills with grass, water is added to make up one-fourth of the volume, the still is capped and sealed with cow dung and fired briskly. Lemongrass oil, which has a clean, sweet citrus smell similar to that of verbena, volatilizes at the temperature of boiling water and contains about 80% citral, used in the manufacture of ionone, an oil only slightly different from orrisroot oil in its violet-citrus fragrance. It is sometimes called lemon verbena oil and sold at a higher price. Lemongrass oil is soluble in 180 proof ethyl alcohol; recovered commercially in vacuo, it has been employed in recent years as a citronella substitute.

Musk Ambrette. Though not a plant, musk ambrette is probably the most important artificial product in the synthetic perfumery industry. Its odor recalls floral notes rather than that of natural musk, but it is popular as a fixative in floral fragrances containing water and is added in the amount of 1%.

Myrrh, *Balsamodendron myrrha*, is a shrubby tree indigenous to the countries bordering the Red Sea. Its gum-resin is collected in tears formed by cuts in the outer bark. A tincture is generally produced by soaking one part powdered myrrh resin in two parts 180 proof alcohol. The almond-cherry fragrance has been used in the holy oil of Jews, and to this day it is a favorite fixative of elegant Middle East perfumes. Myrrh has a very agreeable odor, and on distillation it yields about 5% essential oil. It is extensively used with tuberose, honeysuckle, lavender and sweet-pea scents.

Orange Blossom, *Citrus vulgaris*, is employed to produce aqueous fragrant waters and oil of neroli. Fully expanded fresh blossoms are pinched off with the fingernails, placed in a net sack in a copper pot still and covered with water about one and one-third times the volume of flowers. The still is sealed and a lively fire lit under it. A quarter of the original liquid is collected, and it is generally not cohabated. As a rule of thumb, a quart of distillate is recovered from two pounds (1 kg) of flowers. There are 15 constituents in the volatile oil of orange blossoms, and they vaporize at temperatures between 122° and 240°F (50° and 115°C). Orange blossom water is often used in combination with alcohol and oils for cologne waters because it lends a brightness to lackluster scents. Oil of neroli is largely produced in the south of France by steam distillation; it possesses the sweet odor of fresh orange blossom and adds a delicate bouquet to all floral perfumes. Both neroli and orange blossom petitgrain oils are basic to the perfume trade, and synthetics of both are commonly employed.

Orrisroot, *Iris florentina*, origin of the classic fleur-de-lis, is a fragrant, violet-scented plant, the three-year-old roots of which produce its potent volatile oil. The best quality roots are obtained from the Mediterranean countries where they are widely cultivated in the dry, sun-baked hills. The bulbs arise from eyes that develop on the root. The flowers are sold for ornamental purposes, and after cutting them, the root undergoes rapid change as it becomes richer in its aromatic components. Freshly harvested roots are soaked in water to help remove their skins before they are spread to dry. The starch of orrisroot creates a foam that impedes distillation by boiling water, so most orris oils are extracted by steam or by ether in vacuo distillation. The yield is small, 1% or 2%, but the oil is made up of 15% irone, a violetlike ketone that is an important constituent in perfumes and sachets. Synthetic irone, which combines with lavender and rose/vanillin scents, may be substituted for orrisroot oil; it is said to give a richer depth than the natural product.

I tried to distill plain iris roots and recovered my Witch of the Land Fill, a very garbagey fragrance. I tried to throw it on a 1:00 A.M. visiting tomcat, but I missed completely and for a long time our back steps reeked of the malodorous creation.

Patchouly, *Pogostemon patchouli* is the deep, warm, comfortable fragrance of India that conjures up Kipling, cashmere and gracious brown-tinted people.

For the wind is in the palm trees
And the temple bells they say,
If you've heard the East a-calling
You can't ever stay away
From the warm patchouly odors
And the sunshine and the palm. . . .

Come you back you restless traveler,
Come you back, you've been gone too long.

KIPLINGING IN THE STILLROOM

The upper stems of the sagelike patchouly plant are cut several times a year, partially dried, then piled in heaps to ferment slightly before the leaves are stripped and subjected to distillation. Patchouly is known as one of the finest heavy bouquets. It is used extensively as a fixative in the compounding of oriental perfumes. Scents employing patchouly must mature for at least a month before packaging. Although historically distilled in boiling water, prefermented and dried patchouly is imported by commercial distillers in Singapore and distilled with steam under pressure. The high pressure breaks down inner leaf cells to yield more oil. Scents containing patchouly are said to keep insects out of clothes and carpets. Patchouly oil is one of the finest fixatives for heavy, Oriental-type perfumes, and it is popularly employed in white rose bouquet, a scent of toilet powder and sachets. Bouquet of patchouly is made by blending one part each of oil of patchouly and oil of rose with 100 parts of 180 proof ethyl alcohol. By itself, this *Labiatae* cousin has not a very pleasant odor.

Pennyroyal, *Hedeoma pulegioides*, is a pungent beggar of Virginia waysides pleading to be picked and perked into aromatic pleasures. Cut the plant just before it flowers, allow it to wilt a day or two before distilling the chopped herb in a boiling water still or in a steam still. Collect about one-fourth of the distillate, which begins to condense just as steam starts to come over, and redistill it. Save about one-fourth of the second run. Create an essence by mixing with 160 proof ethyl alcohol, one part distillate to 50 parts alcohol, and add a mite of rose oil for a lilting summer fragrance.

Pine, Balsam, Fir, Hemlock and Larch. Young leaves are commonly distilled into pine needle oil by boiling water. They are employed as room fresheners, in balsam-scented bath oils, in mens' perfumery and in soaps. As a youngster I snatched handfuls of green needles and my grandmother perked them in pine water. This woodsy water was used to combat the ammonia odor of chicken houses. Though hardly a perfume, twice-run pine water made things smell good, and Grandma gave it to her hens to drink; she said it discouraged internal parasites. Pine needles are one of the easiest and most rewarding distillations; most oils pass over between 158° and 185°F (70° and 85°C), and they are soluble in 180 proof alcohol.

Rose. Old time otto, or attar, of roses produced in the Balkan states was distilled from the fragrant *Rosa damascena*, a pink-to-red double rose that grew in florets of 20 blooms. Gathered during May, the roses were brought directly to the still that was a 30-gallon copper boiler with a puffball head and a beak condenser that sloped downward through an oak tub of cold water. The briskly fired still was charged with 25 pounds (12 kg) of petals and two gallons of water, and when the distillate lost scent the spent roses were replaced with fresh blooms. After several rechargings, the rosy rose water was removed from the still, strained and bottled as toilet water. The distillate in the collector was called Persian rose water or otto of roses. Some distillers concentrated the otto, cooled it to 20°F (-6°C), skimmed the top that congealed like egg white and sold it as Bulgarian Oil of Roses.

BULGARIAN ROSE OIL STILL

Distilling roses is pure fun. Hold the plant with one hand, pull the petals from fully opened roses and distill them as quickly as possible after picking. The finest fragrance seems to come over just after water boils, 216°F (102°C). Too long a run weakens and discolors the rose water left in the still.

Last summer a friend brought me a grocery sack of luscious red rose petals; she was traveling and gave me the blooms hoping to come home to bottles of fragrant rose water and a vial of otto. She did not understand that the yield from a couple of pounds of roses is about a scant half teaspoon of otto plus about a cup of colorful rose water. I fired up my trusty apparatus, perked prettily and eagerly poked my nose into the collector. Nothing. Though the distillate sparkled, no fragrance was present. I have since learned that some roses are hybridized for appearance and though eternally beautiful to admire they produce little essence. I have known women like that.

Some people mix the distillate with the boiler water and alcohol, three parts rose to one of 180 proof ethyl alcohol, to create a rose scent. Uncle Frank mixed rose water with lemon verbena and rose geranium to create a light, very feminine scent. He gave me a teensy bottle saying that the boys would chase me when I daubed a bit on my wrists. They did. I was on third base, the batter got a hit and when everyone slid into home I was on the bottom. I heard a boy yell, "What's the smell?" At 8 years old my heart was broken. I also broke my front teeth in that game.

Rose distillates form an important constituent of many compounded fragrances and aromatic waters; in a sense they are the backbone of perfumery. Formulas for the fragrance additive called rose water triple advise 24 pounds (12 kg) rose petals mixed with ten quarts (ten liters) water, and distilled until five quarts (five liters) have been recovered in the collector. This distillate is rerun two more times, each time halving the take until only one and one-fourth (1.25 liters) is saved. All rose fragrances, whether distilled once or several times blend well with verbena, orange blossom, geranium, sandalwood and orrisroot. Some perfumers recommend rose scents for use with all floral tones.

Rose Geranium, *Pelargonium roseum*, is a flowering herb indigenous to South Africa. Parisians first cultivated the rose geranium as a house plant and in 1819 obtained fragrant oils from its leaves. Harvest of the frost-sensitive leaves is done just as the leaves begin to turn yellow and the lemon odor changes to a rich rose scent. Boiling water distillation in a simple pot still has been traditionally employed with rose geranium; vaporization of the iridescent oil begins just before the water boils, at about 210°F (99°C). With the use of steam, however, condensate begins showing at about 167°F (75°C). Immersion in water somehow inhibits early vaporization of the geraniol and citronellol, the aromatic alcohols in this herb oil. Commercial distillers of rose geranium oils employ 150 proof alcohol as a solvent before distillation in vacuo. Rose-geranium oil is most often combined with rose scents, and an old bouquet called Tea Rose combines the two rose fragrances.

TEA ROSE BOUQUET

A cubic centimeter (cc) is the approximate equivalent of a milliliter (ml).

	U.S. Customary	**Metric**
extract of rose	2 teaspoons	10 cc
extract of rose geranium	2 teaspoons	10 cc
orange flower extract	½ teaspoon	2.5 cc
extract of sandalwood	½ teaspoon	2.5 cc
orrisroot tincture	½ teaspoon	2.5 cc
rose water, triple run	2 teaspoons	10 cc
180 proof ethyl alcohol	1 quart	1 liter

Mix the extracts and tincture and allow to stand in a dark, well-stoppered bottle for one month. Add the rose water to the alcohol before combining with the other ingredients. Mature for several weeks before decanting into small bottles.

Sandalwood, *Santalum album*, is cut from trees that grow in the dry mountainous areas of Southeastern Asia. The tree family belongs to a root parasite clan that must be cultivated with other plants. Trees are 25 years old when roots, branches and trunks are harvested. Though some sandalwood is carved into chests and some used in religious rites as incense, most is distilled for its roselike essential oil, santolol. Shredded sandalwood is distilled by steam or by boiling water and yields between 4% and 6% essential oil. Santolol is known for its marked fixative properties, and in addition to its use in Oriental fragrances it is employed in violet, coumarin and musk perfumes.

Spice Bush, *Lindera benzoin*, a North American shrubby tree growing in moist places, yields a pleasant, balsamic, orange-flower fragrance when fresh leaves are distilled with steam. The volatile oils come over between 160° and 270°F (71° and 132°C), and at first the odor is strongly spicy like camphor. With maturation the fragrance changes to a flowery scent, which is useful in bouquets containing lavender.

Vanilla, *Vanilla planifolia*, is a large climbing orchid indigenous to tropical America and cultivated in Java and other warm countries. The glossy green, unripe, odorless, pencillike fruit pods are dried slightly before being piled to start fermentation. After a few days the vanilla pods are dried to a dark brown, and though they smell unpleasant at this stage, enzyme action helps to develop a white powder in their skin wrinkles that contains the fragrant vanillin. Due to the hand work in cultivating and properly maturing vanilla, which includes extraction of the natural oils by alcohol solvents, the use of vanilla pods in perfumery is very limited and has been largely replaced by artificial vanillin. In general vanillin is extracted from wood products and from synthetic chemicals. Vanillin is one of the more important fixatives, modifiers and blenders used by perfumers. It gives body to thin odors, it sweetens harsh scents and combines well with coumarin-based fragrances. One ounce of good vanillin is said to be equivalent to about 40 ounces of vanilla beans. However, I do not believe any substitute can replace the wholesome depth of natural vanilla. If you wish to make your own tincture of vanilla, macerate a sliced, dark, dusted-with-white vanilla bean in a cup of 180 proof ethyl alcohol for a month. Use a tightly stoppered dark bottle and store in a cool place. Because it does not become pleasant-smelling until it is diluted, this tincture must be cut by at least ten volumes for use in scents.

Vetiveria, *Vetiveria zizanoides*, is a tall grass of India from which fragrant mats are woven. The fibrous rootstock, somewhat resembling sandalwood in odor with its camphor-rose overtones, is distilled by steam and the Indian oil is called khuskhus. English vetiver oil is known for its more spicy and spirited fragrance, but distillation is difficult because the boiling point is high.

Violet, *Viola odorata*, is a European species known for its wonderfully fragrant leaves and flowers. Violets do not yield a volatile oil upon aqueous distillation, nor are they completely soluble in alcohol. Years ago violet scents were extracted by enfleurage; the scent-saturated fats were treated with benzoin as a fixative and sold as pomades. Today, violet leaf oil is extracted by solvents, and no violet perfume is said to be perfect without a 3% addition of that slightly earthy fragrance. Natural violet oils are hard to find, which might be a good thing because *Viola* absolutes were said to benumb the olfactory nerves. Contemporary man has enough problems without a numb nose. Synthetic violet extract is made from ionone, naturally found in lemongrass, and it is considered one of the most valuable synthetics. A half-ounce (15 grams) is sufficient to perfume 200 pounds (100 kg) of soap. The odor of diluted ionone recalls orrisroot, and if diluted further with alcohol smells like fresh violets. A curious fact is that when first added to alcohol, the odor of ionone entirely disappears, but develops again in the course of a few days. Undiluted ionone, like true oil of violet, benumbs the olfactory nerves. It is said to hold a "thin" odor, but when compounded with orrisroot, ylang-ylang, sandalwood, bergamot or clary sage oil, ionone gives a full, violet bouquet to perfumes. Synthetic violet (ionone) must not exceed 60% of the floral oils in a fragrance, and it must be matured for at least six months.

Ylang-Ylang, Ilang-Ilang, *Canangium odoratum*, is a small tree indigenous to the Philippine Islands that bears exceedingly sweet flowers that bloom over a long period of the year. Distilled by boiling water or steam, the flower oil is sold to Oriental perfume houses. Ylang-ylang blends well with benzoin fixatives and with vanillin and coumarin scents. My friends who traveled to Southeastern Asia last year brought me a gift of ylang-ylang. Surprisingly, the odor knocked at far-removed memories; scents of Uncle Frank's mysterious room next to the coal bin came back to me. . . . Frank, always bright and merry, had in his youth fallen in love with a Malay girl, and they were wedded by an Asian rite. Deliciously happy, he carried her off to Singapore, dressed her in Western garb, put rings on her fingers and ears and sent her ahead of him to his family in America.

Unfortunately, Frank's kin did not recognize the "heathen ceremony," nor did they try to understand the swarthy, plump girl. They called her Fat Alice, played cruel jokes on her, such as teaching her to call out "toodle-loo" as a salutation and giving her an old corset to wear, telling her it was an outside garment. They also treated her as a servant.

Alice was a mild young woman, strong as a man and eager to please. She gave herself fully to any task and demanded nothing for herself. She cared for Frank's irascible father, she tended Frank's four squalling nephews, she slopped the pigs, washed clothes on a board, cooked and nursed the household for fifteen months while Frank's steamer inched from Sumatra to San Francisco.

Alice was not a promiscuous woman. But one day Frank's brother's wife told her to get out because she had found Alice with the ice man in the barn. It was evident that the unlovely Malaysian girl was lonesome. Her training had been to care for men, to please them, to give them what they asked for. Thoughtless of herself, eternally calm and peaceful in her generosity, Alice returned to the wharf and waited for Frank.

"Where's she at?" Frank demanded when his ship finally docked. . . .

All of this happened years before I was born, but the story was repeated dozens of times. How Frank rummaged through all the corners of town looking for his Alice. How, after two years, he drifted East, a derelict. How he landed in our town, St. Joe, Missouri, and was immediately called Steve.

Fate. People helloed, invited him to share their bucket of beer, the priest, mistaking him for my grandfather, asked him about his plans for the ice cream social. . . . Uncle Frank, though no relative, was a deadringer for my grandfather. At five and forty, his hair was thinning in the same places, his intensely blue eyes, identical in their brimming with bright, good humor, his tall frame, even Frank's grand, thick nose was a duplicate of my grandfather, Steve's.

Granddaddy took Frank under his wing, found him a boarding house and a job and gave him a little room beside the coal bin for his own secret refuge.

When I came to St. Joseph, Uncle Frank's obsession with fragrance, his experiments with the sweet, heavy perfumes of the Orient, lent a mystique to his musky stillroom. The scent of ylang-ylang brought back memories of Frank's merry eyes and of the hurt that must have lain hidden behind them.

The Art of Perfumery

Blending scents is more than creating sweet smells—it is the capturing of memories in a bottle.

When I read of a scientist in England building a bionic nose, a computerized smell sensor whose receptors can distinguish between a whiff of jasmine and the smell of a rose (plus 20 other odors) my brain cells exploded. Think of the magnitude of uses such a sniffer could offer. One could odorize his car keys or his glasses, pick up his portable proboscis and in a jiffy find his lost appendages. Think of an odor receptor's use in detective work, in monitoring who snitched the cookies. But then I thought about capturing memories and somehow my own nose seemed very satisfactory for me.

Although perfumers would hardly refer to teaspoons or cups, measurements are given in U.S. customary kitchen measures and in rounded metric measurements so that quantities will be more meaningful. A cubic centimeter (cc) is the approximate equivalent of a milliliter (ml).

Natural plant oils are rare substances. In today's world, synthetics, or chemical duplicates of volatile plant oils, and carriers such as ethyl alcohol are readily available through private drugstores or perfumery suppliers. Synthetic plant oils may be substituted for most natural volatile oils.

A variety of essential plant oils and fragrant herb plant seeds are available through the Nichols Garden Nursery Catalogue, 1190 North Pacific Hwy, Albany, Oregon 97321.

ESSENCE OF ROSES

	U.S. Customary	**Metric**
oil of rose	5 teaspoons	25 cc
180 proof ethyl alcohol	1 quart	1 liter

Oil of rose is different from most volatile plant oils, thus old-fashioned essence of roses was often turbid. Uncle Frank used to decant his rose fragrances into opaque satin glass containers because he was ashamed that they did not clear. Chemists understand that rose oil contains fats that are viscid. If you make your own essence of roses be prepared for cloudy scent.

EAU DE COLOGNE

	U.S. Customary	Metric
oil of clary sage	1 teaspoon, generous	6 cc
oil of orange blossom	¼ teaspoon	1cc
oil of bergamot	½ teaspoon	2.5 cc
oil of lavender	½ teaspoon	2.5 cc
oil of rosemary	½ teaspoon	2.5 cc
180 proof ethyl alcohol	1 quart	1 liter

Although colognes are commercially distilled in vacuo they may be made by mixing the essential oils, one at a time, into the alcohol (shake after each addition). The product is matured for at least six months in a completely filled, well capped container. One lady perfumer told me that she turned her cologne top to bottom every week during the maturation period. She also kept it in a brown paper sack in the refrigerator during this time.

YLANG-YLANG COLOGNE WATER

	U.S. Customary	Metric
tincture of vanillin	1 teaspoon	5 cc
essence of rose	1 teaspoon	5 cc
oil of neroli	½ teaspoon	2.5 cc
oil of ylang-ylang	½ teaspoon	2.5 cc
180 proof ethyl alcohol	2 cups	500 cc
orange blossom or plain distilled water	2 cups	500 cc

Cologne waters are made up of any combination of tinctures, extracts, essences or essential oils, plus alcohol and water. The plant essences are mixed first with the alcohol, shaken, matured in a filled, tightly capped bottle for a few weeks before being diluted 50-50 with distilled or aromatic water.

Perfumery concoctions using vinegar are not as popular as they used to be because of possible association with skin irritation. It has been recommended that nothing stronger than 10% to 15% acetic acid vinegar be applied to the skin, and hygienists warn that even this vinegar solution be kept far away from the eyes.

MUSTY ROSE ACID PERFUME

	U.S. Customary	Metric
oil of camphor	2½ teaspoons	12.5 cc
oil of rose	1 teaspoon	5 cc
oil of rosemary	¼ teaspoon	1 cc
oil of ylang-ylang	¼ teaspoon	1 cc
vinegar, 10%–15%	1½ quarts	1500 cc

In a dark bottle mix the plant oils with the 10% to 15% (acetic acid strength) vinegar. Shake between each addition of the oils, cap tightly and mature for four or five months in a cool place.

Vinegar Wash

Vinegar washes were made by many stillroom matrons in years past. Soak aromatic herbs, flowers, crushed seeds or slivered roots in common vinegar for a few weeks. Strain or filter and mix with equal amounts of isopropyl alcohol. Splash on as a skin coolant on a hot day.

Aromatic Water

Aromatic waters of yesteryear, and of my stillroom, are made by distilling fresh or dried aromatic plant parts in a boiling water pot still or in a steam still. The still is recharged with new material as many times as there is material. That which is recovered may be redistilled by steam to concentrate the fragrance. These aromatic waters may be made from most fragrant plants; I have distilled everything from spice bush to chestnut flowers and have recovered some enticing waters. Aromatic water may be used just as it comes from the collector or it may be blended with other perfumery ingredients. Formulas that call for dilution by distilled water are naturals for homemade aromatic waters. They should be stored, topped with a skim of 180 proof ethyl alcohol in filled, tightly lidded dark bottles and kept in a cool place.

PERSIAN MUSK COLOGNE WATER

	U.S. Customary	**Metric**
geraniol alcohol	1 cup	250 cc
olibanum extract	2 teaspoons	10 cc
benzoin tincture	1 tablespoon	15 cc
rose geranium oil	¼ teaspoon	1 cc
vetiver oil	¼ teaspoon	1 cc
neroli water	½ cup	125 cc
rose water	½ cup	125 cc

The first five ingredients are mixed one at a time, lidded, shaken and allowed to stand an hour between each addition, after which the mixture is allowed to mature in a tightly stoppered, completely filled, dark bottle for two weeks. The two waters are added and the fragrance is matured a few days before bottling in small containers. Of the four essences, two—the olibanum and benzoin—are fixatives.

Persian Musk was marked on an old-fashioned scent lavalier that I possessed before someone liberated it from me. The chainless pendant opened to disclose cotton scented with a penetrating perfume, which I took to be Persian Musk. It was an aggressive odor.

I came by the lavalier during my short-lived career as a waitress, which lasted one evening. The high-society gathering at the college president's house was jinxed from the start. The honored guest caught his foot in his own trousers and ripped his inseam from stem to stern. The electricity went off during a cloud-burst that flooded the basement. The turkey stopped cooking, the gelatin dissolved, the Baked Alaska sogged and the gravy lumped. However, with numerous stiff upper lips and aperitifs, things swam merrily until one matron, with a bust that would make a Holstein jealous, developed a misunderstanding with her husband as to whether or not their bank would sponsor a college clinic for bad eyes. The lady not only laid down a scented trail when she became excited, but her voice became intimidating. I was ill-at-ease to serve her, and when I removed the salt and pepper I dropped the pepper into the matron's water goblet, causing ice chips to fly. Her low-cut gown caught a fistful of ice. With a gasp she dived in after the chips and deposited them back into the tumbler. Seeing the silver disc in her water glass, I tried to tell her that she had removed her lavalier with the ice, but she waved me away. In the pantry I dried the scented pendant and brought it to her as she was donning her wrap. All at once the lady smiled radiantly, saying that I should keep the piece, that my mishap had cooled a nasty argument. They were going to sponsor an eye clinic, she said, and she was going to be the very first client. Her husband squeezed her waist saying that she finally admitted that her eyesight was failing. She had lost the lavalier, he said, and all the while it was down her front. I grinned, noting to myself that she could have lost her evening purse in the same place and never found it until she undressed. The curious thing was why the lady hadn't smelled the Persian Musk.

ROYAL MIST

Chicken, Alaska; a stranded harlot with a heart as big as all outdoors, with tear-drop diamond ear-bobs, red alligator shoes, all swimming in this heady Royal Mist scent.

	U.S. Customary	**Metric**
neroli oil	1 teaspoon	5 cc
bergamot oil	2½ teaspoons	12 cc
lemon oil	1 teaspoon	5 cc
rosemary oil	¼ teaspoon	1 cc
lavender oil	⅛ teaspoon	.5 cc
orange blossom water	¼ cup (scant)	50 cc
essence of ambergris	1 teaspoon	5 cc
180 proof ethyl alcohol	1 quart	1 liter

Dissolve the oils in the alcohol and add the essence. Macerate seven days and shake frequently. Add two teaspoons (10 cc) of orange blossom water each day for five days, shake daily. Mature at least two months in a dark container before decanting.

SALVIA COLOGNE

Shipboard cabin mate, a neat, nice woman, an undertaker's widow pitifully addicted to drugs. A comforting fragrance.

	U.S. Customary	**Metric**
bergamot oil	4 teaspoons	20 cc
coumarin	1 teaspoon	5 cc
ylang-ylang oil	½ teaspoon	2 cc
rosemary oil	¼ teaspoon	1 cc
lavender oil	¼ teaspoon	1 cc
clary sage oil	¼ teaspoon	1 cc
rose oil	¼ teaspoon	1 cc
musk extract	½ teaspoon, scant	2 cc
180 proof ethyl alcohol	1 quart	1 liter

Mix and macerate for six months in a well-filled, dark bottle.

FLORIDA WATER

A lady friend in Anchorage who was crazy about hats, changed hats several times a day; she was a confirmed hataholic and swore by Florida water. She was buried in a perky red felt and her coffin was scented with this lovely lavender.

	U.S. Customary	Metric
lavender oil	2½ teaspoons	12 cc
bergamot oil	½ teaspoon	2 cc
benzyl acetate, jasmine	½ teaspoon	2 cc
phenyl-ethyl, rose	½ teaspoon	2 cc
vanillin	¼ teaspoon	1 cc
bitter almond extract	⅛ teaspoon	.5 cc
isopropyl alcohol	2 cups	500 cc
distilled rose water	1 quart	1 liter

Macerate the oils and synthetics in the alcohol for one month; add the water, settle a week, filter if needed and bottle.

MANDARIN

Somewhat spicy, sweetly mysterious, Mandarin was Grandma's favorite fragrance.

	U.S. Customary	Metric
ylang-ylang oil	1 teaspoon	5 cc
bergamot oil	1 teaspoon, scant	4 cc
rose oil	¼ teaspoon, generous	1.5 cc
bitter almond extract	½ teaspoon	2 cc
180 proof ethyl alcohol	1 quart	1 liter

Mix and macerate a month before decanting.

Scents crowd the mind with memories. . . .

The scent of bitter almonds, a hermit on the banks of the Tanana in Alaska, a hot, sticky night, mosquitoes humming like a living net around us. Seeing an old-timer hunched on a knoll, unwaving, we drew our boat ashore. The curious man, squatting as if in a trance, did not look up, and coming closer we noticed that his head was dripping with sweetscented almond grease. He neither blinked his glittering eyes nor spoke until my friend produced a bottle. Obviously as attracted to hooch as he was repulsed by soap, with the first swallow the old timer began philosophizing. He had been thinking about his senses, he told us. "I do not feel my sensations like I used to; time robs a man of everything." Drinking every time the bottle was passed, he told us that an infant first feels and hears his mother's pulse;

hearing and feeling are the first senses to "grow weary," he believed. Next the child sees and tastes; man's sight and sense of taste grow dim in ordered succession. Another nip and he informed us that sex urges do not diminish "till them pearly gates is closing on a man's tail." Our informant believed sex to be a sense. "And the sense of smell?" I asked the man as he tipped the bottle for the tenth time. "Smells tell you everything, they lasts and lasts, smells ain't never worn out. They'll bury you for smell if ain't for nothing else." The Alaskan suddenly laughed heartily. He was pleased with himself, pleased with the hot stinking pomade running along his collar. He was enjoying the moment immensely.

That strange man lingers in the corners of my mind with memories of sweet scents, carefree times and thoughts of Uncle Frank. Frank believed that a man's good deeds rose before him like the fragrance of flowers.

6

CHAPTER

Cosmetics

Cosmetics, the ancient Greeks declared, should gladden the senses and aid health and beauty. As a people, the Greeks showed a passionate fondness for physical excellence; where men aspired to artistic and gymnastic dexterity, women were the primary manufacturers of beauty aids, a profession that was immersed in magic and dedicated to celestial deities.

"Here first she bathes, and round her body pours soft oils of fragrance," Homer wrote of a goddess as she prepared to meet her lover. Throughout classical literature the secrets of consummate love seemed to be attained with a boost from cosmetics. Most heroines captivated all who viewed their charms by the use of ointments applied to the face, unguents rubbed on their lower limbs, scented clays as hand soap, pomades for the hair and powdered cuttlefish bone for tooth polish. Though Socrates condemned the practice as effeminate, some males applied luxurious cosmetics to every part of their bodies.

He really bathes
In a large gilded tub, and soaks his feet
And ears in unguent.
His jaws and breasts he rubs with thick palm oil,
Both arms and buttocks with oily extracts of sweet mint.
His brows and hair he powders with marjoram,
His knees and neck he soothes with oil of thyme.

SHORTENED FROM ANTIPHANES, ANTEA.

Loaded as they were with oily cosmetics, the poor ancients probably slid around in their chariots or had a hard time hanging onto their bar stools.

Wealthy Athenian women led sedentary lives, which soon withered their beauty. To make up for lost youth, they painted their faces with white lead, their eyebrows and lids with Egyptian kohl, and their lips and cheeks with red alkanet, the dried root of a *Boraginaceae* that was extracted by immersion in warm olive oil.

Perhaps the grandest creations of all Greece were the hairstyles that both men and women wore elaborately knotted on the crowns of their heads and fastened with jeweled clasps, bright silk, rope or braid. When they reached the age of puberty, men used to cut off their hair and dedicate the shorn locks to a deity, such as Apollo.

Hair coloring was as important as hairdressing. Short of boiling themselves to render them young again, many older men subjected themselves to tortuous hair and whisker dyes. Oxide of lead, quick lime and ground bones were rubbed into a fine powder, mixed with water, pasted onto the hair and beard, held for a day or until the desired tint of brown was obtained. Incautiously applied, the caustic dye acted as a depilatory.

A story is told of a young beauty bewitching an old man until he fell desperately in love with her, but he was rejected. The rich man had his slaves work feverishly with the dye pot all night. When he presented himself the following morning, all dapper in splendid dark hair, the Grecian maid smiled wanly, "How can I give you today that which I refused your father yesterday?"

Several centuries after the glory of Greece began to wane, Rome emerged to rival the Athenians in the adulation of cosmetics. Perhaps their highest achievements in luxury were their magnificent baths. Roman public baths were built by rich men, such as Nero, and were bequeathed to the people for use for a tiny fee. Some men of wealth assigned their estates to support the operation of the baths. Only men were allowed to bathe in baths.

Upon entering a Roman bath, the men undressed, gave their robes to an attendant for spot cleaning or mending, and proceeded naked into a room where crocks of scented oils lined the walls. They rubbed themselves with an unction of their choice, or had a slave do the rubbing, then walked to the cold baths where they scrubbed in icy showers. From there the men hurried into the tepidarium, where they relaxed in sunken pools of sultry water or sat on benches in a fog of steam. After a good soak or sweat, the bathers entered the hot baths, which were built over a furnace fired by wood. There they curried their skin with a pumice paddle, simultaneously dropping scented oils on their bodies. Slaves assisted the bathers at this time. After a rubdown, some bathers wandered to saloons for conversation or to art galleries, enclosed patios or flowered walkways; or they participated in gymnastics or enjoyed a theatrical performance or a lecture; all was part of the Roman bath. The baths were often social and business gathering places, complete with temples to the gods on one side and a local pub on the other. Some baths accommodated 2,000 bathers and contained 1,500 marble or bronze seats.

TEPIDARIUM AT POMPEII

M'lady's bath was less elaborate and not so popular because most wealthy Roman matrons preferred to attend to their own toilet. They each had their own retinue of skilled slaves called *cosmetae*. Making their bodies alluring with unguents and scents, their faces white with chalk, their cheeks rouged with fuchsias, their eyes highlighted with ink and their hair bleached with goat's fat and ashes was the total occupation of some rich women. Oils aromatized with roses, quince blossoms, bitter almonds or narcissus were applied to their bodies, hair, ears and feet. Everything about them—the walls, bedding, dogs, horses, harness—was impregnated with odorous dews. False hair was worn by women and by men who had denuded their scalps with bleaches. At one period, fashion leaders promoted the "invisible peruke." Hair was painted on bald pates with a mixture made of fragrant pomades and stove black.

A ROMAN LADY'S BOUDOIR

Cosmetics have come a long way, only to circle back on themselves over and over again. A lady I knew in California desperately wanted a new husband. She believed that the road to the altar, any altar, was paved with paint. She concocted an eye shadow containing aniline, an oily liquid that colored her lids a deep violet blue and then removed them. Arsenic poisoning is not the way to land a husband.

The first rule in making your cosmetics is be aware of ingredients. Do not substitute unless your pharmacist suggests that you do so. Follow the directions and the amounts stipulated, and if you have the opportunity to meet an organic chemist, make friends. An agreeable chemist and courageous friends who help try out concoctions are friends indeed.

When purchasing ingredients for home cosmetic making, the privately owned, nonchain drug stores generally deal with a variety of suppliers who can more likely meet your needs. Also when purchasing ingredients from a drugstore, the pharmacist may dispense the corresponding approximate equivalent of metric measurements in the apothecary system of measures, or vice versa. A cubic centimeter (cc) is the approximate equivalent of a milliliter (ml); there are about 5 cc (or ml) in a teaspoon.

With the exception of a good candy thermometer, metric measurement equipment and your still (if you plan to distill your own aromatic waters), most tools needed for making simple cosmetics are found in the home. As in making a Chinese dinner, it is wise to study the recipe, make or buy ingredients, lay out equipment and measure, chop, cube, crush or grind anything that can be done ahead of time. Also, have ready a container in which to hold your product. Good relationships are not enhanced by having a night cream ready to jell, pouring it into your husband's coffee mug and having the cup take on the odor of rose geranium forevermore.

If you have a still, perk out a few bottles of distilled water, or buy a jug of it in the drugstore. Cosmetic formulas usually call for it.

When I read in the paper that the average American woman regularly uses over 20 cosmetic products, I scoffed that I did not depend upon that many preparations for my beauty. Then Lewis and I counted 12 regulars with a few now-and-againers. That figure surprised me because I firmly believe that good food, rest, regular exercise and laughter, plus a belief in God that includes a faith in the future, form the rock-bottom basis of health and attractiveness.

Soap and Hair-Care Products

My interest in cosmetics started with hair. When I was a towhead, my grandmother took great pride in my hair; she regularly washed it with soap and rinsed it in lemon water, which she said kept hair light. She rinsed her own hair in vinegar water for a coppery glow. She influenced Granddaddy to rinse his hair in the "Dew of Youth," a rosemary distillate or strained decoction that she believed retarded baldness because of the stimulating qualities of oil of rosemary. She said that good shampoos or hair oil contained rosemary. When Grandma's hair turned gray she rinsed it in store-bought blueing, which she also used in the rinse water of sheets and shirts.

Twice a year I got the mange cure. To my knowledge I have never had mange, but in October and May my head was washed with pine tar soap and doused with mange cure containing pine tar, linseed oil, olive oil and industrial alcohol. I slept in a towel turban, was waked before dawn, subjected to a couple of hot soapy washes and rinses, then dried by the kitchen range before breakfast. I went to school smelling like a pine tree and hated every minute of it.

Looking back, I remember how my grandmother luxuriated in fussing with my straight, silken locks. I know she had visions of a princess; she made me feel like one even though I was all knees, knuckles and nose. Love is very forgiving.

Natural oils keep hair supple, shiny and reduce damaged ends. Years ago in Alaska I saw lush, healthy heads of hair on natives who never heard of shampoo but who eagerly sought a bar of soap. Next to a lemon, the most prized gift in many villages was soap. How the school girls scrubbed with soap. Giggling, laughing as if their whole world were filled with joy, those innocents got high on bubbles.

During winter months in times long past, northern natives washed their hair in urine. This habit is not as repulsive as it may seem when we consider the scarcity of water and the difficulty of heating it. Freshly passed urine contains ammonia, which cuts grease and discourages head lice, a problem of all peoples living in close quarters. Early-day natives rinsed their hair in precious melted snow water, which was soft.

Beauty begins with cleanliness, cleanliness begins with soap, soap begins with fat, and it is all put together with elbow grease.

The difference between fats and oils is temperature. When I explained that to my 4th graders, one young philosopher queried, "Why doesn't my granddaddy melt?" Not knowing, I could only grin and give the boy permission to take the rest of the period to try to seek library answers.

Oils or fats are divided into two classes: fixed and volatile. Fixed oils are generally greasy and leave a lasting spot on paper; volatile oils do not. A cork twisted into the neck of a bottle containing a fixed oil makes no noise; twisted into a volatile oil bottle, the cork squeaks.

Fatty substances are salts that are composed of stearin, margarin and olein; these consist of stearic acid, margaric acid and oleic acid. Stearin and margarin are generally solids at room temperature, whereas olein is liquid. The proportion of olein in a fat determines its liquidity. For example, natural lard contains more olein than beef fat, thus it is softer. Olive oil contains much more olein, and fine shampoos or castile soaps are made of oleins or olive oil.

Traditionally, stearic fatty acids have been mixed with an alkali, such as lye, to form soap. Lye or caustic soda, plus other sodium and potassium alkali products, are strong caustics that should be measured exactly and handled with care. During pioneer times lye was made by leaching hardwood ashes with rainwater. That produced a potash, or potassium lye. Early-day soaps were generally soft, kept in a crock and scooped out when needed.

In today's world, sodium lye, made by an electrolytic processing of common salt, is used in the commercial manufacture of bar soaps.

If lye comes into contact with your skin, splash it off with a flood of cold water, wash with soap and water, rinse well and daub with vinegar or lemon juice. If lye splashes into your eye, flush the eye immediately with cold water and and call a physician. Fear of working with lye is unnecessary, but healthy respect and care are needed.

The alkali or diluted lye of soap does the cleaning. There exudes from human skin and scalp an oily perspiration that catches dust and dries into a film. The alkali of soap combines with the oily skin substance and is soluble in water; thus, with a good scrub and rinse, the greasy dirt washes off.

Shortening Hair Wash

This is a mild, make-it-from scratch soap shampoo that was popular in the 1930s when olive oil and muscle power were cheap. The initial step in soapmaking is to make a lye solution by dissolving lye in water; then the lye and fats are mixed together at a specific temperature and stirred until a chemical reaction, called saponification, takes place. This means that the lye solution and fat mingle to create soap and glycerine. In homemade soap the glycerine is mixed in and not allowed to separate; glycerine helps make soap feel smooth.

Start the day before you plan to make your hair wash by making the lye solution so that it will be cool when you need it. Buy a can of lye, which is usually found in the supermarket cleaning supply area. Do not buy drain cleaners because they often contain other ingredients. You want pure lye, which is caustic soda. Weigh out a half pound (250 g). I use my grandmother's technique of weighing; that is, I put the scales on the cabinet, weigh the container to be used, in this case a half-gallon, canning-type jar, then add to the jar's weight the amount of lye needed. Carefully move the jar to a safe surface and add three cups (750 ml) distilled water.

As the water hits the lye it will heat up, boil and fume; keep your face away from it. With a wooden stick, stir the lye water until it is in solution. Let the lye water cool a bit in a safe place before lidding it securely and letting it stand overnight. Be careful that no child or pet can get into it.

On soapmaking day, dig out a large enamel or stainless steel pan that can be heated over a larger kettle of water to form a big double boiler. Measure into it the following fats: two cups (500 ml) vegetable shortening, four cups (a liter) of olive oil or deodorized castor oil, and one and three-fourths (425 ml) vegetable oil such as cotton seed or soy oil. Heat the fats in the double boiler to 95°F (35°C), and while they are heating, remove the cap from the lye solution bottle. Place the lye bottle on a rack in a second pan of water and heat the lye solution to 95°F (35°C). Temperature is very important. Do not allow it to get higher than 98°F (36°C).

You may need a comrade for the next step, so call an adult who has a good strong arm.

When the ingredients are a cozy 95°F (35°C), remove the containers from the heat, put on long rubber gloves, start stirring the fats with a wooden spoon and slowly pour in a pencil-lead-size stream of lye solution. Keep your face well back. Pouring the lye into the fats in a very fine stream helps saponification. Laying the lye bottle on the rim of the fat kettle and weeping the lye solution into the fat makes pouring easier. Pour slowly. Do not drip lye on yourself. Keep stirring slowly and pouring in a drizzle until the mixture looks like soupy mush and the lye water is all mixed into the fat.

Now stir the shampoo to keep everything in suspension while it cools to room temperature. This is tedious work, but keep going; think of the volume of hair wash you will get; you will not have to buy shampoo for a year, and your product will be healthful for your hair. Think of all the greasy food you will escape by emptying your cupboard of fats. Make up stories, write letters in your head, keep stirring, slowly round and round. Do not look at it.

After about an hour of stirring, just as your shoulder begins to drop off, your product should look glossy and thick and hold its shape momentarily when you drip a spoonful on the top of your kettle of soap. It should look like rich latex paint. Add oil of rosemary, plus other fragrant oils if desired; one tablespoon (15 ml) total. You can create your own fragrance by mixing several essential oils purchased from the drugstore. Make sure they are chemically synthesized or real essential oils; most do not feel greasy. If you prefer an Oriental scent, ylang-ylang is sweet, and mixed with oil of bitter almond, gives off a mysterious musky odor. Gardenia and carnation, both synthetic, lend a sweet clove essence to the hair wash. Cedar and lemon verbena give a woodsy fragrance. I like rosemary and rose geranium for a floral scent with a hint of citrus. The fragrant oils may be mixed ahead but should be capped tightly. As soon as you add the fragrance, mix well, around and around for another five minutes. Pour into containers, lid, and you are done.

Shortening Hair Wash makes about two quarts of concentrated shampoo, and the littlest bit does the job. Like castile or other soap shampoos, this soap will require a vinegar or lemon rinse in the last water.

Hair Oil Conditioners

The simplest conditioner treatment for damaged hair is baby oil rubbed gently into the hair and scalp. After your head is well oiled, hair pin a gallon plastic freezer bag over your hair and let your body heat do the work. Write your mother for 30 minutes or so while you are conditioning, then shampoo your hair, rinse with vinegar or lemon juice in the last rinse, and set.

A product very similar to baby oil may be made by purchasing a pint (500 ml) of mineral oil and adding a teaspoon (5 ml) of lanolin. Buy the anhydrous lanolin because it does not contain water; also insist on deodorized lanolin or your product will smell like a wet sheep. After the lanolin and mineral oil are mixed, you may wish to add a fragrance. To accomplish the best aromatization, the oil should be warmed in a double boiler to about body temperature and about ½ teaspoon (2.5 ml) of the fragrant oil mixed in. Any of the coumarin scents lend a clean fragrance; oil of rose is always nice, but if you use vanillin, the oils will have to be heated to 169°F (76°C) to incorporate it.

Before you aromatize all of your mineral/lanolin oil with sweet scents, you may wish to create a healthful masculine hair oil by scenting part of your hair oil conditioner with oils of pimenta and bay, or sandalwood. A drop of hair oil on each moist palm, rubbed into the scalp, does wonders for unruly manes. Remember when working with essential oils that some absorb oxygen from the air, and unless kept tightly closed, the air can change odors so that they smell like turpentine. Synthetic oils or terpeneless natural oils should be employed when using mineral oil because many natural oils and fixatives are not miscible in clear solution, and no one likes foggy hair oil.

Hair Set

Hair set used to be made with egg white beaten with rose water. It smells good and sets the hair nicely, but it flakes, so if you have dark tresses, egg white hair set makes you look like a speckled hen.

In school one time I swiped some agar agar, diluted it until a jelly formed and set my hair. All worked fine until it got warm. That evening everyone kept asking where the fishy odor came from. As school girls, we all did crazy things. I remember one girl who soaked plantain seeds, *Plantago*, until a clear goo formed, which she employed as a hair set. I have since read of plantain jelly utilized as a set, but my schoolmate boiled her concoction. It dried like mucilage and she missed classes because she was home soaking her head in hot water.

Flax seed jelly has been used for centuries as a hair set. Soak seed, one part to three parts water, overnight. Bring to a boil and simmer for 20 minutes before squeezing through a jellying bag. Dilute if necessary with rose water or scent with oil of rose.

MODERN HAIR SET

	U.S. Customary	**Metric**
rose water	1 cup	250 cc
lecithin	1 teaspoon	5 cc
glycerine	1 tablespoon	15 cc
isopropyl alcohol	2 tablespoons	30 cc
oil of bitter almond	⅛ teaspoon	.5 cc

Request powdered lecithin from your druggist; it is easier to put into solution than granules. With a fork, mix it with a tiny bit of rose water and gradually add more liquid to make a juicy paste. Slowly heat the remaining rose water and stir in the soupy lecithin rose water paste. Stir until the lecithin is fully dissolved, remove from heat, add the remaining ingredients, and when everything is in solution, bottle and cap, allowing to cool before tightening the lid.

BRILLIANTINE

	U.S. Customary	Metric
180 proof ethyl alcohol	1 cup	250 cc
mineral oil	3 cups	750 cc
oil of bay	1 tablespoon	15 cc

Bergamot oil or sandalwood oil may be blended with oil of bay for a more spicy tone. Dissolve the essential oil or oils in alcohol. In a double boiler heat the mineral oil to fever warmth—104°F (40°C)—and mix with the scented alcohol. Decant into bottles, lid tightly and be sure to label. A young friend's mother tried to fry okra in Brilliantine, which the girl had put into vegetable oil bottles.

Petrolatum Pomade

Petroleum distillation produces a residue called petrolatum, (vaselin) which when purified is odorless and acid-free. The substance retains the consistency of lard. For an inexpensive dressing for curly hair, heat in a double boiler to fever heat. Add a few drops of synthetic oil of bay and oil of citronella or pimenta oil for a clove/cinnamon scent.

Creams and Emulsions

Next to soap, creams are considered to be the most important cosmetic in skin care.

Human skin consists of three parts: the deepest layer is made up of subcutaneous cells that gradually change into skin. The second deepest layer is corium, a thick, true skin. The thin outside layer, or epidermis, consists of dead and dying cells. Shedding skin, like a pine tree shedding its needles, is an ongoing process.

The outer skin contains glands that sweat, and the subskin has fat-producing glands that keep the outer layers supple. Hairs poke through the layering like mushrooms through the woodland floor. The primary missions of cleaning, whether by washing with soap or applying creams, is to remove oils that have collected foreign particles, to keep the natural skin glands healthy and to keep body odors at a minimum.

In these days of chemical wonders, the variety of creams—emulsifiers, all-purpose, cold cream, cleansing, vanishing, lubricating and moisturizing—is mind-boggling. Add the lotions, astringents, milks, pastes, masks and softeners, and the do-it-herselfer is defeated before she starts. Ingredients listed on modern packages overwhelm a person with their stearates, palmitates, carbomer thing-a-ma-jigs and numbers. At times it seems that some manufactory mumbles are purposefully duplicit, with the object of pulling the wool over the buyer's eyes. "Them things are put there to mess up the eyeballs," I heard a cosmetic customer complain, and I agreed with her.

Remembering the objectives of cleanliness and health, if you like creams, start with the old-time favorite, equal parts glycerine and distilled rose water. When glycerine is cut half and half with aromatic or distilled water, it has a cleansing, softening and remedial effect on the skin. Never use glycerine straight because it absorbs moisture from the skin and can be an irritant. Some cosmetic authorities recommend 25% glycerine; others suggest 10% each, glycerine and alcohol, combined with 80% rose water. Most drugstores sell glycerine.

Glycerine is soluble in both water and alcohol. If it is heated too hot it will smell like candle wick; it burns like sugar.

Emulsions are products in which particles of one liquid are held suspended in a different liquid. Example: Oil and water do not mix, but with the addition of an emulsifier, the emulsifying agent surrounds the droplets of oil so that they remain suspended in the water. Emulsions are usually milky looking. Milk, seen under a microscope, consists of clear fluid in which the minute butterfat droplets float; by refraction of light, milk appears white.

True emulsions include little, if any, caustic alkalies, but they possess the power to clean skin. Emulsifiers surround the skin oils that have collected dirt, and water or skin cream carries them away. Glycerine is an important component of emulsion creams because it dissolves grease and most cosmetics, and it retards the rancidity of fats.

Years ago, expressed oil of almonds, olive oil or lard, plus glycerine, were the basis of cleansing emulsions. Today most formulas substitute mineral oil and chemical emulsifiers in the manufacture of skin creams.

GLYCERINE SKIN CREAM

	U.S. Customary	**Metric**
glycerine	½ cup	125 cc
mineral oil	1 cup	250 cc
chamomile water	¾ cup	175 cc
spermaceti wax	¼ cup, scant	50 cc
beeswax	1 tablespoon	15 cc
oil of rose geranium	½ teaspoon	2 cc

If you wish to distill your own chamomile water, pick the daisylike chamomile flowers, fill a steam still three-fourths full, add an inch or so of water in the bottom, heat until boiling, connect the tubes and fire briskly. Almost immediately volatile oils will begin to rise because chamomile vaporizes between 104° and 300°F (40° and 148°C). There will be a pretty blue film on the distillate.

Years ago in its natural state, spermaceti was found in the skull cavities of whales and dolphins; today it is synthetically prepared and often called by different names. Consult your pharmacist, who can recommend a substitute if synthetic spermaceti is not available. The properties of spermaceti stand midway between beeswax and paraffin in hardness. It melts at about 104°F (40°C), and when candles were a source of illumination, spermaceti and beeswax were considered the finest ingredients from which candles could be made.

To make Glycerine Skin Cream, slowly heat the slivered beeswax and granulated spermaceti in a double boiler to about 158°F (70°C). When liquid, add the mineral oil.

In a second container premix the glycerine and distilled chamomile water; add the oil and wax mixture to the watery solution, and stir vigorously. Using a blender was recommended to me; a lady told me to trickle the scented water/glycerine mixture into the warmed oils, which had been put into the blender, and to have the blender on high when pouring in the water/glycerine in a thin stream. "It comes out like mayonnaise," she said. In either case: oil into water, or water into oil, with vigorous stirring; continue whipping until the mixture cools to fever heat and carefully fold in the oil of rose geranium. Pour into well-stoppered bottles and keep in a cool place. This skin cream will be about the consistency of egg white; if you want a thicker cream, increase the beeswax by about a tablespoon (15 cc). You will need low, flat jars for the heavier cream. This cream may be applied as a conditioner or as a cleanser. Camomile flowers are said to have a soothing disinfectant effect on the skin.

COLD CREAM

	U.S. Customary	**Metric**
beeswax	5 teaspoons	25 cc
spermaceti wax	5 teaspoons	25 cc
mineral oil	1 cup, scant	220 cc
lanolin, anhydrous	1 teaspoon	5 cc
orange blossom water	½ cup, generous	150 cc
rose geranium oil	½ teaspoon	2.5 cc

In a double boiler, gently heat the slivered beeswax and granulated spermaceti wax until 158°F (70°C) is reached. When liquid, add the premixed mineral and lanolin oils. Keep the heat at the same temperature, stirring all the while, until the mixture clears. Turn off fire.

In a separate pan have the orange blossom water heated to 95°F (35°C). Slowly, but with continuous stirring, pour the warmed water in a thin stream into the cleared oils. Stir constantly until the temperature drops to 104°F (40°C) and add the rose geranium oil. Continue stirring until the mixture cools before pouring the cold cream into containers with lids. This cold cream is definitely greasy. A little thick liquid poured into the palm of one hand, applied with the fingers of the other hand and removed with tissue leaves the skin clean, smooth and refreshingly soft. I have not made this cold cream, but the lady who gave me the recipe (and said that she always blends watery substances into oils, not vice versa) let me sample her product; it was really very lovely.

Working once in a large drugstore, I was assigned to candy because I am not partial to sweets. A coworker with wide-set lavender eyes who was in charge of cosmetic sales used to come visit my station when customer traffic lulled. She loved malt balls, and while we chatted in front of the malt case, she explained that in her native Hungary she had been a perfumer working with cosmetics. She said that she could tell a cosmetic maker by her well-developed upper torso. They used to stir their cosmetics by hand, she said. She added that it was important to mix cosmetics gently around and around with, once-in-a-while, a swipe down the middle of the pan. She talked expressively, using her hands and her eyes and all of her upper torso. She was built like a cosmetic mixer.

Hand Cream

Over a double boiler melt two ounces by weight (60 g) anhydrous lanolin that has been deodorized. When melted, remove from heat and mix in two teaspoons (10 ml) glycerine and six tablespoons (90 ml) deodorized castor oil. Beat well or blend until a uniform creamy consistency is reached and the mixture has cooled to body temperature. Add one tablespoon (15 ml) isopropyl alcohol to which a few drops of benzoic acid and a half teaspoon (2 ml) lavender oil or other fragrance has been added. The benzoic acid counters rancidity in fats and gives perfumes holding power. This is an excellent skin emollient.

Elm Hand Jelly

Though I would not cut down a slippery elm, *Ulmus rubra*, for this skin softener, if I found a broken elm tree I would chop off the bark, sliver the sticky inner bark into a pan and soak it in an equal amount of warmed rose water, elder blossom water or plain distilled water. The mucilaginous gum contains a gelatin with healing properties. The material must not be stirred. When well soaked, in about four days, remove the glutinous mass by slithering the thumb and forefinger along each side of the wood slivers and dropping the jelly into a bowl. The slightly aromatic lotion prevents chapping and combats dryness. Store in a cool place because Elm Jelly will ferment if left at room temperature for more than a week.

Thirty years ago in Alaska I flew light planes. On bright winter days my habit was to jog to the airport for a midday spin around Seward's peaks. One brilliant noon, as I passed an abandoned airport shed, I heard movement. I thought that a raven had entered the broken window and was trapped, so without hesitation I pushed inward on the door and stepped back expecting the bird to fly out. Nothing happened. With a shiver of uncertainty I looked inside. At first all was darkness, then a motionless form outlined itself against cracks of light on the far wall. I am no hero; I ran to the hangar for help.

The Eskimo woman we led out of that miserable hut was thin and pale as tallow. The marble-black eyes that gazed out from below her parka ruff glistened. It was apparent that she was poorly fed. Cardboard stuck out from the soles of her mukluks and her fur-lined parka was in shreds. We took the youngish woman to the hospital where she was cleaned, examined, fed and put to bed. When I stopped to see her that evening, she had left. She had spoken no word, just put on her clothes and, unnoticed, slipped out. Authorities were notified but no one had seen her.

The weather turned surly and a week passed before I returned to the airport. Though no sound was discernible, I had the feeling as I approached the shed that someone was within. This time I called out, knocked and pushed the door open, all the while talking cheerily as if my best friend were inside. The little native woman was squatting down; I handed her a square of cake, my dessert from lunch that I was bringing to Tiny, a sweet-toothed airplane mechanic.

Eating with obvious pleasure, licking her fingers, the Eskimo smiled and we became friends. Her son was in the tuberculosis sanitorium, she said, and she had mooched her way from Wainwright, in the Arctic, to be close by when, "Time come for died."

Damon did not die. Hattie, his mother, was hired as a ward helper at the sanitorium, and within a year both returned home.

That winter I received a gift from Wainwright. It was wrapped in a greasy paper that smelled so wonderfully of wood smoke that a pilot friend who delivered the package licked his chops for smoked salmon. His face fell like a pancake when I pulled out moosehide mittens.

The huge suede-tanned mittens, which had beaded backs and Hudson blanket insides, with wolf hair cuffs that reached halfway to my elbows, were works of art; but most surprising were the scented and oiled, kidlike, baby-sealskin fitted gloves that I discovered loose inside each mitten.

Scented gloves! Shades of Shakespeare, they had been popular in England under Charles I.

The following year I flew north to visit Hattie and Damon. Melted caribou fat scented with herb and flowers of ledum or Labrador tea, *Ledum decumbens* or *L. groelandicum*, had been used in the soaking and stretching process Hattie used in tanning the sealskin. The leather had become as supple as mink and it smelled grand.

For years when flying, I wore them inside my moosehide mittens. After removing them, my hands smelled spicy as the northern tundra. The oiled leather prevented chapping and they most certainly did gladden my senses and aid my health and beauty. What wondrous people, those Eskimos!

As I waved goodbye to Hattie and Damon, the bright-eyed boy ran forward to my window. "We see into each other's heart," he called. His shy, grinning mother had prompted him.

During a recent trip to Maine I smelled ledum tea, and I smiled remembering Hattie's words.

Borax and Soda Vanishing Cream

Vanishing creams are so-called because they disappear when rubbed into the skin. Basically they consist of stearic acid and water. Natural stearic acid is obtained from tallow and other fats by foaming the fat with lime and treating it with steam at high pressure before sterilizing and purifying the product. The finest quality, odor-free stearic acid, melting at about 133°F (56°C), should be used in cosmetics.

	U.S. Customary	Metric
stearic acid	4⅓ oz. by weight	130 g
borax	2 oz. by weight	60 g
sodium bicarbonate	½ oz. by weight	15 g
rose water	3 cups	750 cc
glycerine	¼ cup	50 cc
rose geranium oil	2 teaspoons	10 cc
patchouli oil	¼ teaspoon	1 cc

In a double boiler melt the stearic acid. In a second pan, heat the rose water and glycerine. Make a paste of the borax and sodium bicarbonate with a little of the water, and when it is in solution, mix it into the water and glycerine and bring to a boil. Slowly pour the melted stearin into the boiling solution. Stir. Continue to simmer until the mixture gelatinizes. Cool to 104°F (40°C) and add the perfumes. Bottle and cap securely. If the ingredients separate after being bottled it means that the fats were not completely emulsified and the product will have to be shaken before use. This does not affect its virtues.

Skin Bracers

"Skin bracers not only make a man feel good, they improve his disposition," a man who made his own bracer told me when I used to ride with him on the bus to Washington. Proud as Punch, he said that he melted lard, added a sliced vanilla bean, emptied the liquid into a widemouth, half-gallon Mason jar with a tight lid, and kept it warm in the pilot-light-heated oven for several days. Next he heated the jar of vanilla lard in hot water until it became very liquid. To it he added a fifth of clear rum. He shook it each day and put it back into the oven for several days. He then refrigerated his concoction. When the lard had congealed he scooped it out, together with the vanilla bean, and bottled his rum bracer. He smelled good, but in all honesty his odor was more like a newly baked cake. He offered to bring me a bottle of bracer for Lewis. I declined, but Lewis said I should have taken a sample; he wanted to taste it.

Witch Hazel

An astringent skin bracer to which Prohibition winos were said to have been addicted was witch hazel. I have not nipped it, but distilled hazel water tastes terrible. Witch hazel is regarded as an excellent additive to skin lotions of all kinds because it is mildly antiseptic and nonirritating; authorities state that there is no better astringent.

Witch hazel is made from the twigs and leaves of the shrubby tree *Hamamelis virginiana*. Fill the copper pot still boiler three-fourths full of hazel leaves, add water to cover, secure the head and boil the still vigorously. A strongly scented water will come over first, and if you can generate enough heat, a greenish fatty oil will distill and float on the top of the distillate. Hazel is a slow-distilling oil; maximum vaporization occurs at about 257°F (125°C). Although steam distillation produces a strong-smelling aqueous solution, the end product does not seem to possess the same oils that are recovered through boiling water distillation. Commercial houses distill hazel leaves and stems in vacuo. To make the product known in the drug trade as Liquor Hamamelidis, or witch hazel, 140 proof ethyl alcohol is added at the ratio of one part alcohol to five parts distilled hazel water, or about 17% alcohol.

A COMMERCIAL DRY STEAM STILL

Hazel trees, which generally grow along the banks of rivers, greatly resemble alder. In addition to the lopsided hazel leaf base, another distinguishing feature of the hazel is that it blooms in November. When other woodland life lies dormant, *H. virginiana* breathes signs of spring with bright yellow squiggly flowers. To me it seems as if the hazel tree was frustrated with its scrubby status in the tree kingdom, so it made up its woody mind to bloom when it pleased. I see it as a plant that invented an imaginary world in which freedom to choose when and how it flowered became a reality. I wonder if alder trees growing with hazel on the river bank ever raise their flora hackles and cry, "Discrimination!" when they see their neighbor being allowed to blossom so prettily in late autumn. There exists a danger that the privilege of freedom to choose may be confused with discrimination.

Because of the tannic acid found in hazel leaves, the distillate of hazel contracts tissues and lessens irritation and bleeding from small blood vessels. Thus witch hazel may be employed in aftershave and cleansing lotions.

LIQUID CLEANSING LOTION

	U.S. Customary	**Metric**
hazel water	1⅓ cups	330 cc
borax powder	2 tablespoons	30 ml
140 proof isopropyl alcohol	½, cup scant	100 cc
orrisroot oil	¼ teaspoon	1 cc

With a spatula on a marble slab or a smooth surface, mash the borax into a fine powder and transfer to a cup. Add a little hazel water, mix to make a smooth paste, and add more water until the paste is in solution. Mix the other ingredients in a larger container, add the borax solution, swish around a few times and bottle. This lotion is drying for oily skin. If the borax refuses to go into solution, heat a little of the water to encourage it to dissolve.

One time while driving along a country lane I spied a broken slab of marble amid debris in a ditch. The old tombstone facing had no markings, it poked up as if loath to associate with the trash along the roadside. Lewis has been better about stopping ever since we discovered an unopened six pack by the side of the the road. We rescued the slab, scrubbed it and put it to use as a mashing surface.

Genuine Bay Rum

If you live in the south or in the West Indies, liberate a batch of young bay leaves from the shrubby bay tree, *Pimenta acris*. Fill your steam still three-fourths full of fresh to slightly wilted leaves, add a mixture of one part 180-proof ethyl alcohol or clear rum to five parts water to just cover the leaves. Fire your still and distill over a gentle heat until you have collected about one half of the original liquid, or until the distillate runs watery. Vapors will begin to rise at about 158°F (70°C). The still should be kept perking but not heated much over 176°F (80°C). The condensate will be the pleasantly clove/peppery/lemon–scented Bay Rum of commerce, which may be used as a refreshing skin tonic as well as a hair rub. Old books state that Bay Rum stimulates the growth of hair, but pharmacists with whom I talked shook heads.

Synthetic Bay Rum is made by mixing one tablespoon (15 cc) oil of bay (not oil of sweet bay) into two cups (500 cc) of fever-heated 180-proof ethyl alcohol. This alcohol is highly combustible and should be warmed over hot water. Do not heat alcohol over an open flame. When the bay oil and alcohol are cozy to the touch, remove from the heat. Add one-fourth teaspoon (1 cc) oil of pimenta (allspice) and one-fourth teaspoon (1 cc) oil of neroli (orange). When the alcohol is scented to your pleasure, dilute with two cups (500 cc) distilled water. Orange blossom water may be substituted if you have it. This tonic is often sold as Bay Rum, but the fragrance and astringency is less smooth than that of the genuine product according to Uncle Frank.

Uncle Frank made the genuine thing by infusing wilted green bay leaves in the residue from West Indian rum barrels that had been sent to the Missouri vinegar works via river barge. A friend who worked for the company somehow sent him the leaves. Uncle Frank briefly infused the bay leaves in the fermenting rum dregs before adding water and distilling the crude spirit into an aromatic skin bracer. He was reputed to have sold his product to local apothecary shops for 25 cents a bottle. I pasted labels for a penny a dozen until Grandma found out about it. Those who knew about such things said Uncle Frank's Bay Rum was the finest spirit west of the Mississippi. I now suspect that very few of my neatly labeled bottles ended up among the toiletries of St. Joe.

ELDER BLOSSOM BEAUTY FRESHENER

This is an astringent that makes your face feel freshly kissed.

	U.S. Customary	**Metric**
glycerine	1 tablespoon	15 cc
elder flower water	1 cup	250 cc
borax	1¼ teaspoons	6 cc
140 proof isopropyl alcohol	2 tablespoons	30 cc
lemongrass oil	⅛ teaspoon	0.5 cc

Distilled elder blossom water was called *Aqua sambuci* in medieval times, and it was used plain as a gentle skin stimulant and mild astringent. To make the skin freshener, the required distilled elder water, is mixed a mite at a time into the borax to form a juicy paste. The remaining water is heated to 95°F (35°C) and gradually added to the borax paste. Stir until clear. Cool to room temperature, add the glycerine and alcohol, into which the fragrance has been dissolved. Mix well, bottle, cork tightly and allow the freshener to mature for two weeks before use.

A young woman I met said that she mixes a few spoonfuls of elder blossom water with homemade yogurt and allows it to stand a day or two before spreading it on her face as a freckle remover. She applies it several times during her half-day treatment period and "washes her freckles away." I smiled to myself thinking that her freckles looked cute. Faith is a great thing.

Assorted Cosmetics

RICE DUSTING POWDER

	U.S. Customary	**Metric**
rice starch or flour	2 cups	500 ml
cornstarch	10 tablespoons	150 ml
talcum	⅓ cup	80 ml
zinc stearate	7 teaspoons	35 ml
powdered orrisroot	5 tablespoons	75 ml

Mix all of the ingredients in a large bowl using a folding motion, as you would "fold in" cake batter ingredients. After the powders are well mixed, sift through a No. 200 mesh or through a double nylon stocking bag. To form the bag, pull one nylon stocking into its mate so that you have a double thickness of nylon mesh. Tie off the foot. Into this fist-size bag empty the powder mixture. With two fingers in a milking motion, gently sift the rice powder into a gallon plastic bag. Work as much as possible within the plastic in order to keep down the dust.

Zinc stearate is a light powder with astringent and hygienic properties.

Grandma used to make rice powder without fragrance. She added the scent by sprinkling an ink blotter with her favorite perfume, fitting the blotter into the bottom of her powder box, filling the box with powder and placing a second perfumed blotter into the lid. After a couple of weeks in her underwear drawer, the powder box was put on her dressing table for use. Grandma did not believe in wasting anything, not even smell; her underclothes were made fragrant with the maturing powder.

TALCUM POWDER

	U.S. Customary	Metric
talc	4 cups	1 liter
cornstarch	6 tablespoons	90 ml
zinc oxide powder	1 tablespoon	15 ml
coumarin and vanillin	1 teaspoon, total	5 ml

Place a little talc on a smooth surface, make a pocket in the powder and add the fragrant oils. Mash with a spatula. Gradually mix in more talc until all signs of moisture have been absorbed. Fold in the other ingredients, mix and sift through a No. 2 mesh or through a double nylon stocking bag before boxing.

I have been told that coumarin may be obtained as a powder, a synthetic called salicylic aldehyde, but I have not used it.

I distilled sweet clover, *Melilotus*, blooms and leaves, because they smelled like new mown hay, which is coumarin scent. I filled my steam still three-fourths full of flowers, added a little water, fired the boiler and no sooner had steam started bubbling into the water under the flowers than fragrant vapors started weeping into the condenser. The oils of coumarin-containing substances volatilize at about 158°F (70°C). The distillate was not as fragrant as the plant, nevertheless I wet a blotter with it and inserted it into a sample of talc. Result: brick-hard talcum with no smell.

DRYING POWDER

An unscented hygenic drying powder for your juiciest angel, your bed wetter's bottom.

	U.S. Customary	Metric
zinc oxide powder	10 tablespoons	150 ml
cornstarch	1½ pounds	750 g
powdered boric acid	1¼ cup	300 ml

Like zinc stearate, zinc oxide powder is used to deter minor skin irritation and inflammation.

Sachet

Sachets were originally made by pulverizing flowers, herbs and aromatic parts of completely dried plants. They were sold in fancy baglets or paper envelopes to layer among clothes. I tried sachet-making once and discovered that all aromatics do not dry to be sweet-smelling; my violets smelled like the barnyard, and dried lily of the valley turned into dried sauerkraut. Mints, rose petals, lavender and some cleavers retain their fragrance; powdered orrisroot helps hold the scent. Mix the powdered herbs and rasped woods or roots, then cheat a little: Add a drop or two of fragrant oils. Intimately mix in the oils and put the powder into a double-tied plastic bag for maturation before sewing spoonsful into fanciful pouches.

Creme Sachet may be made by heating spermaceti wax with the tiniest bit of beeswax in a double boiler to about 158°F (70°C), cool to 104°F (40°C) and stir in your favorite fragrant oil. Mix well and pour into a small jar with a tight lid. As a nonspill scent for pocketbook or travel bag, a daub applied to wrist or inner knee creates a jiffy fragrance.

Tooth Powder

One of the finest tooth powders may be made by mixing salt and baking soda, half and half. Less salt may be used but some should be included because it is a tooth abrasive and has an astringent action on the gums. Premix the salt-soda, cover tightly and sprinkle a little on your brush for an inexpensive and healthy cleaning job. If you wish to use a whitener from time to time, place hydrogen peroxide, 3% antiseptic (not bleach for hair) next to your salt-soda mix. Pour a little powder in your palm, moisten it with peroxide and scoop the paste onto your brush. Authorities state that you should not use peroxide with every brushing but that the salt-soda powder is a good cleanser and antiseptic. Teeth should be brushed downward or upward away from the gums and rinsed thoroughly.

Tooth Polish

Before fancy pastes were developed, tooth polishes were sold in jars, and the cleaner was scooped out with wooden applicators, and spread on the teeth. A finger was used to polish the teeth. Later, lead foil tubes and brushes made of natural bristles were employed for cleaning the teeth. Celluloid and plastics replaced natural bristles and wooden applicators. An 1800s tooth polish formula calls for castile soap powder, French chalk talcum, orrisroot powder, sugar, distilled water, and oil of cloves or peppermint.

Mouthwash

When making products for the mouth, a primary mission is to neutralize acids and deter bacterial growth. All fermentation systems are go in the mouth; the mouth has warmth, moisture, air and the remnants of food. Fermenting itself is not harmful, but like wine left too long before bottling, acid forms. Acids create bad breath, they eat into tooth enamel, and gums become infected. The best mouth rinse is plain salt water. Mix one-half teaspoon (2 ml) in a half cup water and swish it around your teeth after brushing. Rinse with cold water.

Another mouthwash may be made by mashing enough regular aspirin tablets to make a scant one-fourth cup (50 ml) of aspirin powder. Gradually mix the powder into a little clear, 80 proof rum, to make a paste of the aspirin. Finally mix in about a pint (500 ml) of rum. Dilute by adding a pint (500 ml) of distilled water, and flavor with one-half teaspoon (3 ml) oil of peppermint. If you use commercial flavoring extract, make sure it is for internal use, and mix it with the rum before adding the plain distilled water. The salicylates of aspirin soothe inflamed tissues and are antiseptic, the rum carrier preserves, and peppermint water flavors this surprisingly refreshing, sweet and clean-feeling mouthwash.

Lotion for Sore Feet

A handsome woman from southern Maryland, who by her own words used to walk like a duck because of callouses, told me about this lotion when I had a toe with the screaming mee-mees and I could not keep up with my pick-your-own, strawberry-picking companions. What a kind soul. Imagine, a person stopping to help a stranger with sore feet.

Mash a handful of aspirin to make about one-fourth cup (50 ml) of powder and gradually mix in a cup (250 ml) isopropyl alcohol and a pint (500 ml) of glycerine. Dilute with a quart (liter) of distilled water and shake well before using. "You make yourself some of my lotion, warm it, and when you sets yourself down, soak them critters. Dry him good and rub him with a plain aspirin pill. Your callous will d'solve hisself." This product may be returned to its container and used over and over, my co-picker told me.

I tried it and after several treatments my corn disappeared. The soaking refreshes tired feet, too.

FOOT POWDER

	U.S. Customary	Metric
corn or rice starch or flour	3 cups	750 ml
powdered aspirin	¼ cup	50 ml

Mash aspirin to make a powder and mix the powder with corn or rice starch or flour. Dust your feet after bathing, sprinkle a little foot powder into your shoes to retard foot odor and deter infections. Salicylates in aspirin check excessive sweating.

Preparations for the feet always remind me of Eliann Dora. She was called a scryer. She was not an ordinary crystal ball gazer; she insisted that her clients wash their feet with brandy before each reading. Mrs. Dora believed that the fate of man was determined by the character of his feet. A man's feet, reinforced by head, hands and heart, held the reins of his destiny. Thus she directed customers to hold her crystal ball, which she ceremoniously placed on the floor between the soles of their feet.

One dull Saturday about a half-hour before sunset, the air suddenly became limpid and an emerald shimmer gleamed across the untroubled wavelets of Sitka Sound. Meda Harmon and I were on the dock awaiting the shoreboat to take us back to Edgecumbe Island where we worked in the Indian school. As we stood on the wet planking, the ozone-laden air intensified until the heavens turned steel blue and warm moist odors rose from the tide flats below. Town noises were hushed. At once the resonant voice of a gull rang out, "El-eee-ann!" Simultaneously a sweep of wings became audible, and the bird, head extended, flew from behind the dark fish packing plant in a path directly overhead. Its unwavering eye seemed to cast a spell as it passed. "El-ee-ann! El-ee-ann!"

Whether moved by the bird's eye or the evening's magic glow, Meda decided to visit Eliann for a reading and I tagged along. Meda was a believer in the psychic sciences. She read signs in everything from the shape of a slab of liver served at lunch to the curl of smoke from her cigarette. She even forecast events from pages of a Bible randomly opened.

Meda also was intrigued with numbers . . . 268, 269, 270. She counted our steps to Eliann's shack, which clung to a mud hill at the far end of town. "Seventy-one," Meda cried triumphantly as we approached a rickety stair. "The number one means action, a good sign."

We entered Eliann's room without knocking, for the door was ajar and blood-red, glittering beads and candles invited us. Once inside I noted that a clock under the belly of a rearing horse and a half-eaten sandwich crowded either side of Eliann's crystal ball, which resided on her chiffonier. Her bed, a sagging remnant of its vanished youth, was covered with an Indian blanket.

The crystal gazer was an outgoing, neatly dressed fat woman with a narrow forehead, protruding eyes, broad lips and jowls that hung like solid brown leather from each side of her mouth. Her friendliness exuded confidence.

Meda, a gaunt nurse in her forties who had a long nose, small gray eyes and a healthy sheen to her curly bobbed hair, smiled merrily as she gave Eliann $3, then sat and removed her footwear. She had visited Eliann before; she had known her in Fairbanks and had sought her counsel in Juneau. In fact, on several occasions after Meda had lost her husband Kreg to a "black-haired witch in Nome," the two had been buddies on the Yukon River steamer. Since her abandonment, Meda had repeatedly traveled throughout the northland to plead with her roving spouse.

After her foot wash, Meda sponged her feet with brown liquid that smelled like alcohol. At that time Eliann, who had greeted Meda with sincere warmth, turned to her crystal ball, and gazing into it for a long moment, ceremoniously wiped it with brandy and with near rapture placed the ball between Meda's feet, which were reclining before her on a dark leathern rug. In a kneeling posture before the crystal, the scryer placed both of her hands over the ball and feet. Holding them thus, she rocked side to side. No one said a word. Eliann's face assumed a look of extreme concentration, and slowly she moved her hands from the glass ball and pressed each palm on Meda's feet, moving them together and turning them so that the crystal ball was held between the arches. The lines of her feet were magnified in the glass.

I was standing near the open door, and to me the ball appeared to gradually become dark, dark red. The gazer concentrated on the crystal, and in a constrained voice said, "I see a stone, a stone frog. He leaps everywhere. He, you, a face. You have been searching for the wrong man, you've been in love with the wrong man. A frog. Trouble. A fall. You cannot catch a frog, he is slime."

"But here there are white clouds, I see them moving up, a good sign. Yes. A financial gain. Watch for a dark man who comes from the East. He has a, a . . . His arm is pinched, no, his arm, I cannot see. His arm, right arm, good luck. His hand, he carries something, a paper. I see the letters K. H. Oh!"

With the exclamation Eliann withdrew her palms from Meda's feet and covered her face. I saw her shiver.

"What is it? What did you see?" asked Meda moving forward on her chair. "What happened?"

Eliann looked thoughtful as she raised herself. "He squashed the frog. He covered the frog with paper. It is no more. . . . There is no more."

"Here, wash the slime off your feet," the fortune teller said abruptly. "Here, use the foot wash. Rinse, wash away the sorrow. There's no more to see. You go now."

The brown woman appeared exhausted as she pointed us to the door.

"Foot wash!" I scoffed cheerily as Meda and I hurried along the dark street toward the boat landing. "Foot wash! In my opinion there's no fortune to be seen in a glass ball unless it's filled with gold nuggets. It's all foot wash," I said firmly.

"But Eliann is a good actress," I admitted after a while because Meda was not making a peep. "She's good-hearted and she gives you your money's worth," I added.

As we neared the dock Meda began defending the crystal gazer. "Eliann can look into the future," she said. "When signs are right she can summon up images that inspire her predictions. She was right, you know. Kreg is a frog, a slimy, no good frog."

Three weeks later while on a preorganized tour of an abandoned gold mine at Chichakof, a half-day's boat ride from Sitka, Meda fell through a rotted boardwalk plank. I noted that the caretaker who ran forward to help her had but one arm. Meda was shaken but unhurt, and as we stood around waiting for her to compose herself, the mine guard, a middle-aged, clean-shaven man with agreeable features, told us that he had lost his arm in Nome. "A donkey engine flywheel caught me; we were sifting sand for gold there on the beach." Later, noting Meda's surname, he asked her if she knew a barge hand named Kreg Harmon. Meda blushed and looking up from the walkway where she was sitting, her eyes filled with moisture; they took on the blue of the sky. She searched the mine keeper's face as she asked apprehensively, "Kreg?" At that moment she appeared to be as fragile as a Dresden doll.

"Kreg Harmon ain't around any more," the man volunteered. "He had a line on a barge directing it into the landing here about a month ago. He slipped and it crushed him. You knew him?"

"I knew him," Meda mumbled. "I knew Kreg Harmon." With the words she bawled like a helpless child.

The embarrassed caretaker uttered condolences then took her arm to help her to the mine office. Insurance papers, he said, and you could see him melt with Meda's tears.

"Destiny," Meda confided to me a few hours later on the boat trip back to Sitka. "According to Eliann's reading my destiny is being fulfilled. Kreg was a frog. As it turns out he may have been a frog worth more dead than alive." A chord of bitterness vibrated in her voice. She sucked in sea air and scornfully expelled it; she ground the word, frog, through her teeth and would not be distracted. But when our tour boat rounded Cape Edgecumbe and the sun unexpectedly crowned the mountain with luminescent clouds, Meda stood transfixed. "There's good luck coming," she declared with a sideways glance at me. "Like Eliann said, white clouds are rising. Bill Newton, the mine guard, is going to help me fix up the papers; he has one arm, you saw that; and my fall, and financial gain. All is being fulfilled according to the signs, and," Meda Harmon giggled like a schoolgirl, "I think I am falling in love."

"Whoa, wait a minute, a coincidence," I interrupted. "I heard Eliann, she said a dark man. Bill Newton is as pale as a parsnip, and he came from Nome, that's north. You're talking foot wash." I said. But Meda was not listening to me, she had turned and was studying our wake. "Foot wash!" I repeated half angrily.

Corny as it sounds, Meda, who had collected a bundle from her late husband's insurance, and Bill were married within the year. I went to the wedding. "Still think it was foot wash?" Meda whispered to me after she had tossed and I missed her bouquet.

7

CHAPTER

Materia Medica

There is a kind of a man who's a cold fish,
And a kind of a man who lies,
To be both is to be contemptible,
A worse creature God never devised.

There are braggarts who blow hard and swagger,
Old maids of the masculine sex,
Though often abused and more often used,
They can be fearless and brave as the best.

KIPLINGING IN THE STILLROOM

Dr. Roger was a sparely built, sharp-faced man with a watchful expression and a despicable little mustache. Most people did not like him. There was no warmth about the man; he could be blunt as a cob, vain as a cock, and sometimes, on the commercial end of his business, he was not quite truthful. But the doctor was respected as a physician. Professionally he was above reproach and he treated a full schedule of patients in his tightly run Alaskan clinic.

Energetic as a titmouse, he possessed an insatiable scientific bent; he dissected and distilled anything that arrived in his laboratory. He was of the old school, a man who tested everything himself and who prepared his own medicines. Far into each night his lab light was lit as over a fire he distilled waters, spirits or volatile oils to extract or evaporate juices of plants or vinous substances. He created acids and alkaline salt for pills, ointments or potable prescriptions. Dr. Roger was a man apart, a craftsman and pharmacist as well as a physician.

After supper I often went over to his lab, which was located in the furnace room, and although he rarely spoke a civil word, I enjoyed working with his laboratory contraptions, especially his still. I had come from a long line of distillers and it was like going home, like a journey back into Uncle Frank's secret room in our cellar.

The doctor never smiled; only on a long, down-hill ski slope did his square teeth become visible. He was a ski addict and had come north after snow.

By day I worked for Dr. Roger as an office clerk. He had hired me because I was a ski nut; all of his medical staff were ski enthusiasts. By mutual consent and whenever possible, we all closed shop to go skiing on a glacier or on highland corn ice; even Mt. McKinley's eternal whiteness knew our tracks. We worked hard and we played hard, and we loved it.

Returning from a ski weekend one spring, our train stopped at Nenana where the river ice was momentarily expected to go out. The ice had been metered to register the exact moment of break-up and hundreds of people, who had wagered on the precise break-up time, gathered along the bank watching the seemingly motionless river ice draw the metering line tighter and tighter. From below the river's edge the water growled as if irritated by the crowd's excited clamor. Above, a caustic wind whistled along the frozen waterway northward toward its mother Yukon. On one hand, the scene exuded riotous carnival pleasures; on the other, the agony of a new season's birth soberly depicted the miracle of life.

Watching from the railroad track, we five medics witnessed a sudden, explosive volley undulate across the valley below us. A wave of riverside spectators fell in unison, as if shot. The church bell bonged, birds flew up. At the same moment we saw the whole center nerve-ridge of river ice rear like a killer whale before shimmying and crashing straight into the water. We saw the towering violence heave boxcar-sized pieces of ice before it as the mammoth slosh wave bore down on the shore below.

People along the bank scrambled this way and that, and when the cumbersome watery ice surged across the land and receded, miraculously no one had been hurt. The chatter and calling and laughter resumed almost immediately as people clustered around the judges stand to await the announcement of the winning break-up time. In the river, huge floes slowly passed in review, and as we watched, the water became choked with shattered, water-logged ice. The winter's product was leaving; the merry holiday pitch of voices welcomed spring.

Suddenly a silence, a sort of unanimous gasp swept over the crowd; everyone stared at a slow-moving floe on the river surface. A figure, having climbed onto a block of ice about 40 feet from shore, lay prostrate, whimpering like a dog.

"Save him!"

"He'll drown!" screamed a woman.

"Save him," repeated a shout as the floe was carried slowly past the viewing stand.

Abruptly, without saying a word, Dr. Roger raced down the bank, and leaping, skithering, slipping over the jumbled ice, he reached the stranded person, seized him by the coat and, pulling, tugging, lifting, flung himself back over the tumbling floes to shore, where he dropped the man-child, a slow-witted Nenana native.

So unexpected, so swiftly executed, so great the risk of the doctor's act, the crowd was paralyzed with astonishment. Several moments passed; it was only when Dr. Roger turned and started for the train that people could grasp what they had witnessed. At once everyone began to shout praises. In reply the doctor waved indifferently, "Only fools and wise men hesitate," he threw out. The people shook their heads and exchanged glances of amazement.

Medical doctors come in all shapes, sizes, colors and fabrics; Dr. Roger was an original. I have worked for numerous doctors in addition to him, and I have come to believe that self-diagnosis and self-treatment of physical ailments are risky. The only safe course is to seek competent medical care when a disorder or symptom of a disorder appears. This chapter is not intended as a treatment guide but as an illumination of the curious paths our ancestors explored in the still-room. The distillation of medicinal herbs formed the basis of medieval medicine; today organic and inorganic chemistry dominate pharmaceutical procedures. The roots of both medicine and pharmacy may be traced to the still.

The Use of Plants in Medicine

The concept of distilling plant parts to extract their medicinal properties probably originated in the mosaic of Arab cultures that dominated the Middle East between the 6th and 10th centuries when Arab scholars of mathematics, astronomy and chemistry emerged as the leaders of the intellectual world. Muhammadanism encouraged scientific study as an ideal that would banish all misery and disease from the world. As scholars delved into the makeup of plants, botany and medicine evolved. The medical properties of herbs, human pathology and herbal treatments of ailments were explored.

Unfortunately, for several hundred years after 1000 A.D., scientific ideals were sidetracked onto metaphysical sidings. Even as Muhammad taught the doctrine of one God being the source of creation and judgment, belief in the miraculous infiltrated Middle East learning. Lifetimes were spent trying to distill a magical substance that would cure all ills and change rock into gold. The Philosopher's Stone—that is, health and wealth—was sought in the distillates of all earthly materials. As alchemy thrived, medical progress stagnated.

Concurrent with the development of the philosophy aimed at transmuting base metals into gold and discovering a universal cure-all, Arab thinkers devised the "science of the stars:" astrology. Astrology suggested that each star and planet of the universe influenced all things on earth. Plants and their active properties took on astrological meanings; today many occult signs are Arabic symbols that attribute magic to plants.

While Islamic tradition emphasized that peace was the concept of consensus in the community, covetous neighbors eyed Islam. The Arabs were rich; as middlemen they controlled the spice trade and much of the world's gold. Within a short period, Persian invaders began unraveling Arab unity. The aggressors took advantage of the Arab peoples' belief in mysticism, and using the zodiac, they predicted wars, plagues and the doomed destinies of southern Muhammadan leaders. The invaders backed up their predictions with military action until Arab leadership was routed and Arab scholars retreated to remote monasteries. Cloistered, man tends to inbreed his own thinking; thus Middle Eastern men kept their stills hot trying to distill rock into gold and plant essences into a cure-all.

Crusaders, inflamed by Holy Land desecration, marched hot-eyed into the area, discovered the Muhammadan monastics with their scientific and metaphysical beliefs and returned to Europe with their helmets full of both ideas. Shortly thereafter, Europeans found "barbarians" barking at their own home fires. With Roman control kaput, travel became dangerous and interaction of thought was limited. During those dark days ecclesiastics began to document the Crusaders' findings. Delving backward, collecting manuscripts that returning churchmen had brought to Europe, the Christian monastics gathered and recorded documents that began with Dioscorides, the Greek physician of the lst century. Little by little, century by century, they traced medicinal learning through Arab scholarship, down to Parkinson of the 1600s, and a theory of healing the sick with herbal concoctions emerged.

In medieval villages, the still evolved as a tool to extract a few home remedies; in church centers, which often were havens for the destitute sick as well as for indigent scholars, distillation became an accepted technique for the recovery of medicinals. Even though some European scientists barked up the alchemist's tree, medicine advanced. With the advent of printed books, distillation methods and apparatus designs spread through the known world. In addition to the distillation of wine to recover spiritus vin, as brandy was called, distillers learned to produce "burnt waters"—strong, rectified (redistilled) alcohol. These rectified spirits were used as a solvent of plant oils. By redistilling the alcohol that had been saturated with volatile plant oils, scientists were able to separate the component parts of a plant oil to isolate its medicinal properties.

During these Middle Ages, as universities evolved separately from monastic study arenas, the art of healing became separated from pharmacy. Apothecary shops pursued the distillation of herbal medicine, but at the same time they kept their toe in the door of alchemy. The distillers' involvement with the alchemy, astrology and mysticism that had surrounded Arab science influenced physicians to turn away from the use of herbs. However, some plants held their place in medicine. For example, the powdered leaves and seeds of foxglove, *Digitalis purpurea*, were prescribed as a heart stimulant. The roots and leaves of nightshades, such as *Solanum nigrum*, and *Atropa belladonna*, produce a poisonous alkaloid called belladonna that paralyzes sensory nerve endings in the skin and relieves pain. The dried seed capsule juice of an oriental poppy, called opium, was used by physicians as the most efficacious drug administered for the relief of pain. The leaves of henbane, *Hyoscyamus niger*, contain an alkaloid that was used as a sedative in cases of delirium, mania and pain in inflammatory conditions.

With the exception of using a relatively few plant materials, medical doctors turned to other substances to provide treatment for the sick, and herbal medicine developed its own discipline. Pharmacists of the 16th and 17th century incorporated both inorganic chemical concoctions and herbal distillates in their realm of expertise.

Different Types of Early Medicinals

It must be emphasized that experimenting with the distillation of drugs is dangerous. Herbal extracts are often too potent for the untrained distiller to explore, so study and extreme caution is advised. The still was, and is, a tool of medicine. From the standpoint of history, the distillation of herbs has provided a very real step forward in the advancement of modern scientific work, but self-treatment with herbs is not recommended without professional advice.

Though medicine traditionally employs the apothecary system of measures, this overview of stills in medicine for the sake of consistency and ease of understanding gives measurements in U.S. customary measuring units and their approximate metric equivalents.

Laxative

It seems that medieval physicians, and Dr. Roger, believed the art of medicine began with a laxative. Doctors of old started treating patients by prescribing boiled flaxseed in water or whey, straining and drinking a dram "to purge without inconvenience." As soon as Dr. Roger's patients arrived, he gave them a Hustler, as we staffers called his concoction. The laxative really was pleasant and quite mild; I made up many jars of it during my years with the clinic. Boil equal portions of barley and prunes in double the volume of water; drop in a few chopped figs and a stick of licorice root and cook until you have a dark mass. Strain, bottle, refrigerate and take a small glass when needed.

Lime Water

A second basic treatment was lime water, sometimes used as a basis for medicinal infusions or decoctions, or as a medicament. Out-of-doors, a half-pound (250 g) of fresh quick lime was poured into a large glazed jug with a stopper. Little by little a gallon of water was poured on the lime. The reason for adding water gradually was that if poured all at once, the lime would form a muddy paste. The mixer was advised to stand back so as not to breathe the fumes as the lime bubbled furiously. When the ebullition was over, the lime water was stirred with a stick, corked and allowed to settle for two days. It was filtered through straw-covered mesh or coffee filter paper and stored in a cool, dark place. A glazed or glass jug with a tight lid was used as a container.

Lime water was used plain as a mouthwash for spongy gums. It was painted on mucous surfaces or swabbed on open sores as a cleanser. The old books say that a decoction made by boiling comfrey root, garlic, rosemary and apple peelings in lime water was administered as a blood-building tonic. A small glass was taken daily. In the stomach, lime water was said to neutralize acid in cases of hyperacidity, thus relieving pain and gas; after absorption it counteracted the formation of acid in the tissues, thus it was prescribed in rheumatic conditions.

Aqua Calcisi was an old tonic and stimulant, a wineglassful taken daily for improving the blood. A handful each of shaved sassafras root and sliced licorice rope candy were simmered in two quarts of lime water for several hours and allowed to cool before being strained and stored in a cold place.

Distilled Curables

Years ago, watery distillates impregnated with plant "virtues" were called aromatic waters. They were the basis of many medieval medicines. Simple boiling water distillation was employed: The vegetable material was placed on a rack in a pot still boiler, covered with water, subjected to heat, and the vapors were condensed. Aromatic waters generally consisted of the first fourth of the amount of the original liquid. Boiling the still longer usually resulted in a more watery distillate.

Medieval distillers learned that some plant properties remained in the boiler water and that the heavier residues sank to the bottom of the still. They knew that heavier plant substances, when exposed to the sun or dry heat, caused water to evaporate and residues to become thick. Often these residues were reduced to tarlike oils, acids or salts, and they were employed in healing.

Spiritus waters, made from fermented plant or fruit juices to form an immature wine, were popular natural curatives prescribed during the Middle Ages. These nonaged, mildly alcoholic liquids were given in women's complaints and for colic in children.

VENDOR OF MEDICINALS AND FRAGRANCES

Spiritus vin was made by distilling fully matured wine or beer and concentrating or rectifying the product by repeated distillations. These spirits of wine or grain brews were considered a man's tonic. They were clear and nearly tasteless. In addition to being administered unaged, as a medicine, spirits were employed as solvents of medicinal plants. Most essential plant oils, or "active principles," as pharmacists call volatile plant parts, unite with various proof alcohols and dissolve in alcohol. Plant parts that resisted solution in boiling water distillation were sometimes dissolved in spiritus vin. The medicinal plant was infused in the alcohol for a specific time, the infusion was strained, diluted and then administered.

Aromatic spiritus waters were made by steeping spices or herbs in spiritus vin for four or five days in a closed container, then straining, diluting by half and distilling. Aqua Melissae, prescribed internally to cool fever, externally to cleanse sores and as a rub to ease the pain of gout, is an example of an aromatic spiritus water. Two pounds (1 kg) of fresh lemon balm flowers, one-half (125 ml) grated lemon peel, one cup (250 ml) crushed coriander seeds, one-fourth cup (60 ml) bruised cinnamon and two tablespoons (30 ml) dried angelica roots were macerated in a gallon of spiritus vin for a week. The plant parts were strained, the aromatic alcohol was mixed with an equal quantity of water and distilled with gentle heat, "just sufficient to raise the spirit." It was perked until five pints came over.

Resins, plant sap that has been hardened by air, age or heat, generally dissolve in alcohol. If they resist solution in high-proof alcohol, they may be heated with fats to become oily fluids. Medieval physicians sometimes prescribed resins, which were heated in fats and their vapors inhaled, as lung or throat curatives.

Plant gums, though not easily liquefied, were employed in stillroom medicine as poison antidotes or as demulcents to soothe inflamed membranes. Gums usually occur in roots and stems of plants, such as in comfrey, *Symphytum officinale*, or acacia, *Acacia senegal* (gum arabic). Gums are generally softened by macerating in hot water until they become glutinous.

Plant Acids

In addition to distilled aromatic waters, spiritus waters, aromatic spiritus waters, spiritus vin, resins and gums, natural plant acids, such as acetic acid, malic acid, tannic acid and tartaric acid, were administered to the sick during medieval times.

Crystals of tartar may be obtained by boiling, half and half, green and ripe grapes until mushy, squeezing the pulp through a cloth and allowing the juice to cool overnight. Crystals of tartar shoot to the sides of the vessel, and they may be scraped off, collected and dried for use.

Cream of tartar was made by boiling the grape tartar crystals in two times their weight of water and repeatedly skimming off the chalky substance that formed on the surface of the water. This chalky tartar was then dried in the sun or by low heat. Taken as a mild laxative, 1½ teaspoons (8 ml) in a large glass of water, tartar water was a tart, cooling medicine popular in European homes. In cases of mucous in the throat a tartar water gargle was prescribed.

Tannic acid (tannin) is found in the leaves and bark of many trees and shrubs, such as oak and sumac. Tannin is soluble in water, alcohol or glycerine. A bitter astringent solution was made by boiling bark in water, straining the brown liquid and often concentrating it by simmering overnight before settling and bottling. During stillroom days, tannic water was used to harden tissues and lessen the secretions of mucous membranes with which it came in contact. When applied externally it was used to treat burns and to slow bleeding from small blood vessels. A strong tea containing tannin was recommended as an emergency antidote for alkaloid poisoning, such as from codeine, morphine and strychnine.

Malic acid is found in the skins of apples, peaches and other fruits. It is soluble in water and extracted through decoction. Years ago malic acid was prescribed for ladies with the blahs: those of sedentary habits who complained of listlessness and digestive problems. It is interesting to note that malic acid is a component of the Krebs cycle, which provides energy in living organisms. An apple a day or a malic acid decoction would seem to be a case of early-day observations that have been sustained by science.

Acetic acid is produced when yeast-fermented substances have passed their alcoholic stage and are attacked by *aceti* bacteria to create vinegar. Natural vinegars were said to aid digestion. When their purpose was antiseptic, that is to check gastric fermentation, a diluted solution was given before meals; when acetic acid was administered to supply acid to the stomach, it was given with or after meals. Acetic acid solution also was given to allay thirst, and it was believed to relieve corpulency.

It has long been known that acids can injure teeth; if taken internally, the acids were well diluted, and the mouth was rinsed with water.

Salts

Salts were another common medicine of yesteryear. When two salt solutions of different concentration meet, even when separated by a membrane, the stronger solution withdraws fluid from the weaker until they become the same strength. (This happens when you make salt pickles.) In animals the process is called osmosis. When salt was taken into the stomach or intestine, it withdrew water from the blood and surrounding tissues and created a need for the ingestion of more water. Early physicians believed that water purged bad humors; thus they prescribed salt to increase water intake and urine flow.

Salts of the alkalies were known as neutralizing agents. The simplest saline materials were produced by burning vegetable matter and infusing the ashes in water. After a week of frequent stirrings in order to saturate the water with ash salts, the water was carefully poured off through a cloth and evaporated in shallow pans over slow heat. Dilute ash salts were taken for bone pain, a remedy similar to but far more ancient than Dr. Roger's lime water.

Complex salts, called alkalies, though very often caustic, were understood and carefully administered by medieval physicians. Salt medicines of stillroom days included caustic alkalies for the removal of warts; dilute solutions of carbonate salts as cathartics; and potassium bicarbonate and sodium bicarbonate given in a large volume of water as an antacid.

Caustic soda (sodium hydroxide), caustic potash (potassium hydroxide), and sodium carbonate (washing soda), all of which are poison, were used externally only. All salts are potentially poisonous; even preparations as simple as sodium bicarbonate should be used cautiously. Indigestion may be a temporary manifestation of stress, it could be a signal of peptic ulcer, it could be motion sickness, or it could be an indication of a serious disorder. Dr. Roger lectured each patient who admitted to taking regular soda products.

Distilling Medicinal Herbs

In general, stillroom wives restricted their medicinals to the watery distillates, infusions or decoctions of familiar garden herbs and plants of the fence rows. Nowadays, with the proliferation of toxic sprays, odd-ball seeds and escaped flora of exotic lineage, warning flags regarding the use of botanicals must be raised. In addition to basic rules about positive identification of any plant before employing it for any reason, it is advisable to go to the library and study the properties of a plant in question. A good herbal will point out known plant constituents, plus their virtues or hazards.

Medieval recipes for curables have come down through history by the crooked path of trial and error. Fortunately, the errors had a self-destruct mechanism and wiped themselves out. Though botanical medicine cannot cover the wide spectrum of pharmaceutical medicine, the healing properties of some plant products should not be discredited.

Herbal distillates are an exciting pursuit. Besides pots, crocks, stirring sticks, a hank of cloth, a candy thermometer, and small, dark, well-lidded containers, only a glass steam still or a trusty copper boiling water pot still is needed. Distilling is pure fun.

Alder, *Alnus glutinosa*, bark or catkins were gathered in earliest spring and infused into an astringent tea, taken internally to reduce pain in the joints and quiet the stomach. Cold application of a decoction or alder bark distillate was used to reduce swelling of sprains and to treat wounds and ulcerations. It was believed to be antiseptic.

To distill alder water, strip bark and lay the thin pieces loosely on a rack in a copper pot still, cover with water and boil the still vigorously. Collect about one-half of the original liquid as it comes through the condensing tube. The active principles of alder volatilize at about 218°F (103°C), and they are soluble in 160 proof alcohol. Alder roots are reported to be strongly emetic and are not recommended without professional guidance.

In Alaska, the train from the north arrived with its weekly cargo of orphans destined to live in the Seward Native Boarding School, and Nick Davidoolik, who had the face of pug dog, disembarked, wide-eyed and hypercharged. At age eight he was a remnant of military presence and of tuberculosis that had killed his young mother and her family. Nick got one eyeful of the drab school, and with the sure, unruffled steps of an arctic fox, he paced from the compound. Before dark he was discovered in a railroad culvert, brought back, comforted, fed and tucked into bed. The barracks held 30 such frightened children. Nick's cot was placed next to the dormitory mother's apartment. He did not cry, nor wet the bed, nor retreat into a catatonic trance like other shocked youngsters. Nick became a bubbly defiant favorite of the school.

Although he spoke English with clipped uncertainty and could not read, he was placed in my third grade class because he was tall for his age and the primary grades were overflowing with other orphans. Brown as alder, bright, he was a joy in the classroom; nevertheless he was given the name "Whereze," from the principal's habit of entering the room asking "Where is he?" Nick was a runaway champion. When apprehended, often after several nights in Seward's deltaland wind, his black eyes danced triumphantly; he loved the game. At school socials Nick danced, squatting and leaping as he had seen men dance in his mother's Indian village. At movies he shouted warnings to the hero and laughed when Indians were gunned down.

One week in spring when slush and birds songs promised better days, Nick evaporated. The river was out of its banks, school furnaces were flooded, sewage backed up, and the chicken house burned down. Everyone was too busy to look for Whereze. I helped pluck, clean and can 158 asphyxiated chickens, and the following day I walked along the railroad tracks to get the stench of burned feathers out of my head. The track was the only high land leading away from the school compound. About a mile north of town, past the trestle that shimmied with the force of the river below, I spotted smoke in the woods, and following boot tracks, I came to a tie cutter's derelict shed. Nick was inside. His shin from knee to ankle was laid open and he was administering to himself by sponging brown-stained water on the wound.

"Tea," he said, his eyes looking like onyx against his pale yellow skin. At the same time he handed me a handful of alder and willow switches and directed me to gather more from the saplings outside the door. "Gift to all the world," he grinned. He was barely nine years old, yet he remembered the shrub as a native healing substance.

As I piggy-backed Nick to the school infirmary, he showed me where he had hurt his leg as he had jumped off the track into an old snow bank when the train had passed. Something in the snow had cut his leg.

The infirmary nurse, who had worked in St. Michael on the Yukon, said she knew of the northern people's faith in alder and willow. She said boiled bark was a useful astringent, and as a weak tea, an alder infusion was beneficial in cases of nervousness, headaches and rheumatism. I watched her paint the wound red and bind Nick's leg with gauze before handing him a lollipop. "Two medicines are better than one," Nick quipped as he hobbled, clownlike, out into the sunshine; and waving his lollipop he called over his shoulder, "Three are better still." My heart lifted seeing his antics, but I wondered what he sought when he ran away.

Barberry, *Berberis vulgaris*, stem and root bark are shaved, washed and laid to dry in a shady place. When brittle, store in paper. Administer in the form of a bitter infusion, the old books state; barberry bark tea was given for dyspepsia. Larger doeses were used as a mild purgative. Barberry bark distillate was given, one-half teaspoon (2 ml) in a glass of water, in cases of jaundice during the Middle Ages. The raw bark is yellowish, and I wonder if the color influenced its use as a jaundice medicine. The distillate also was said to be excellent as a gargle for sore mouth and as a cleanser for skin eruptions.

Bedstraw or Cleavers, *Galum aparine*, was gathered in June and the whole herb dried in a shady, airy place until it became as fragrant as fresh hay. *Galium* comes from the Greek word, *gala*, milk, which indicates the plant's traditional use as a color and curdling agent of milk in cheesemaking. My herb mentor, Ruth Smith, suggested that the milk-curdling bedstraw may have been *G. verum*, which has yellow flowers and grows in Europe. The dried herb of both cousins has been esteemed as a blood purifier and as a quietive for hysteria. It was infused in boiling water, four teaspoons (20 ml) cleavers to two cups (500 ml) water. A rural Virginia family takes cleavers tea three times a day for head colds. The mountain lady told me that it makes a person sleep like a kitten.

The dried flowering plant tips have also been macerated in vinegar and prescribed as a remedy for epilepsy. The active principle apparently is formed during drying. Cleavers distillate is made in a steam still with gentle heat, the volatile properties coming over at about 208°F (97°C). Taken at bedtime, a spoonful in a glass of diluted and slightly sweetened vinegar water was said to be pleasant and restful to the nerves.

Bergamot, *Monarda didyma*, should not be confused with the Italian bergamot citrus tree whose expressed oil is used in perfumery. This bee balm plant, also known as Oswego tea, is cut while in bloom, dried in an airy place and stored in paper. It yields a high degree of thymol, which has been used in medicine as a decongestant since the 1700s. This native North american plant provided an old-timey remedy for sinus problems. A bunch of bergamot was simmered in a flat pan on the back of the kitchen range, a large towel was thrown over the head of the person being treated and also draped over the steaming pan of bergamot. The patient was instructed to inhale the vapors. It was said to clear the passages. Care had to be taken not to breathe the steam for too long nor spill the simmering decoction. If you wish to try this inhalant, you may wish to move the steaming bergamot pan into the sink and do your draping and inhaling there. Incidentally, this steam treatment opens the pores of facial skin and readies it for a good cleansing.

The volatile oil obtained from the whole bergamot herb was used as an antiseptic in dressing wounds during the Civil War. It distills at about 185°F (85°C).

Blackberry, *Rubus villosus*. Bark is harvested from old canes and roots in the fall and dried. A decoction of this highly tannic material was an old remedy for dysentery and diarrhea. As a tonic for pregnant women, the roots were distilled in a boiling water still and about a teaspoonful of the distillate was taken in sugar water, blackberry juice or wine.

An English woman told me a charm to cure rheumatism. While drinking a decoction she called Bramble Bush Tea, she directed that the imbiber say, "Evils will rue/The evils they do,/May the devils in my bones,/Drill into *you*!" The person was supposed to face in the direction of their enemies while encanting the pox. Unfortunately, the lady looked straight at me when she playfully repeated the hex. Her words have not helped my long-bone grumpies one bit.

Boneset, *Eupatorium perfoliatum*, was regarded as a remedy for old-timey catarrh or flu. The leaves and flowers of this coarse perennial were gathered in September; they were wilted, chopped, laid loosely on a rack in a copper pot still, sprinkled over with a handful of salt, covered with water, and the still was boiled briskly. The vaporization point of the principle constituent, eupatorin, is reported to be 244°F (117°C). About one-half of the original liquid was collected as it came through the condenser. Boneset water was taken, one-half teaspoon (2.5 ml) in a scant one-quarter cup (50 ml) of warm port or elderberry wine every half hour for four doses, all the while the patient remained well covered in bed. The medicine was said to induce sweating, sleep and subsequent relief of flu symptoms. My mountain friend said that chill is the main danger, so the patient must be kept well covered. The lady distills her boneset water each fall and premixes it with wine to have it on hand for flu. She also uses a wineglassful for her aged mother's daily tonic and gives some to her husband when his muscles ache. She told me that some people simply infuse the chopped fresh boneset leaves in whiskey and take it as a fever medicine during the winter.

Borage, *Borago officinalis*, leaves and tops are picked just as they are coming into bloom. They are slivered and infused for a few hours in cold wine. The alcohol dissolves some of the potassium salts in the herb, even though they cannot be tasted. Borage wine is said to make the mind glow. Fresh borage is ornamental and faintly cucumberlike in salads, and it does seem to lift the spirit; but then I tell Lewis it might be the good company.

Borage leaves have been distilled into a watery extract since stillroom times. Using steam, insert the herb with a small amount of water, distill until about one-half of the original liquid has condensed. Borage water, when taken in a large glass of cold water, was said to promote kidney action and reduce fever. Borage contains potassium and calcium. If you are watching salt intake, you may want to look into this herb.

Burdock, *Arctium lappa* and *A. minus*, roots dug in the fall of the plant's first year and seeds from the second-year plant have been used as an ancient blood purifier. Decoctions of root and seeds were boiled in water to cover until concentrated by half; a wineglassful was taken several times a day. If a person swallows the stuff, he would be purified alright; he would have no intestinal fortitude left to do anything naughty; it tastes awful. An infusion of the leaves taken as a tea was said to relieve indigestion. Employed externally as a poultice, burdock leaves were believed to reduce inflamed bruises. A distillate of burdock seeds used to be made by putting the dried seeds, sans burrs, in a small boiling water still, covering them with water and boiling them gently until about one-half of the original liquid was collected. The distillate was later run a second time and one-fourth of the liquid was recovered. This fluid extract from the seeds, said to contain the active principle, lappin, was taken, 20 drops, two times a day, in cold water. It was said to sooth severe nervous complaints. The roots of burdock were said to be antiscorbutic.

Butternut, *Juglans cinerea*. A man I met at a cocktail party crowded up to our group to say that he had crowned himself King of Constipation until he found that white walnut, the inner butternut tree bark, infused with vodka, "knocked me off the throne." He embarrassed me and I stammered that it must be pretty powerful stuff. With the zeal of a new convert the man jumped into details of his old difficulty and subsequent cure. He said he chopped away some horny bark from near the base of a butternut tree and chipped thin quills of the inner, dark brown substance for his medicine. Packing a bottle with the chips, he filled the bottle with 80 proof vodka, lidded it and put it in a dark, cool place for a month. The tincture turns dark as a walnut, and he said that he takes about two tablespoons

(30 ml) a day. "Now I am King of the Throne, a regular king," he blustered. I shuddered and sought other company. Since then I have read about the gentle oily properties of butternut bark. The old books say that the oil is also a vermifuge and is recommended for ulcers and syphilis. I have not tested the tincture.

Camphor, *Cinnamomum camphora*, is a large evergreen tree indigenous to East Asia that must be 40 years old before planters will take leaves and twigs for use in distillation. Camphor oil, mixed 50 to 1 with olive oil, is used as a liniment for soothing local rheumatism, sprains and neuralgia. Like sassafras, which is camphoraceous, camphor has a strong odor, a biting taste, and if taken internally in large doses, is said to have an effect on the motor reflexes of the body. Physicians of the Middle Ages relied greatly on camphor distillates for use as an excitant in cases of heart failure or as a rub to stimulate circulation.

Carrot, *Daucus carota*. It is relatively recently that carotin, a vitamin A source, has been isolated in carrots, but the herb of this popular root vegetable has been recognized since antiquity as valuable in the treatment of liver, kidney and bladder complaints. The whole carrot herb was coarsely chopped and laid loosely in a net bag, which was submerged in water to cover in a copper pot still. The still was heated and brought almost to a boil. The active properties come over between 160° and 200°F (70° and 93°C), and about one-half of the original liquid was saved. One-half to a scant teaspoon (2 to 4 ml) of the fluid was taken morning and night as a diuretic, used to increase the volume of urine. The intake of large amounts of drinking water was also prescribed.

When excessive amounts of water are taken, caution must be used that the food intake, particularly vitamins, is adequate because water flushes the system.

In the 1930s our neighbor vowed that carrot tea, brewed from the whole young plant and drunk morning and night, cured her low back, "kidney" trouble.

Cayenne, *Capsicum*. The name cayenne is reported to have been taken from the Greek word meaning "to bite." I have steam distilled dried cayenne into a brilliantly clear liquid, but strangely that firey-looking distillate was not pungent. Old pharmaceuticals instruct dissolving one tablespoon (15 ml) of hot pepper distillate in a cup of rose water as a gargle for sore throats. Other herbalists say that capsicum decoctions or distillates added to tonics are unequalled for stimulating the appetite and warding off diseases. Maybe the Bloody Mary originated with that idea. My peppery hot pepper friend mashes a few cayenne peppers into vodka for a counterfeit Pertsovka, which he vows is a great tonic.

Cedar, *Juniperus virginiana*. In Appalachia, a mixture of the red cedar fruits and twigs is boiled and inhaled as a treatment for bronchitis, and the decotion is applied externally as a skin irritant. *J. virginiana* should not be confused with *J. communis* of Europe, from which an aqueous distillate, rob of juniper, is made from crushed berries and employed as a diuretic. Taken internally, red cedar may be toxic.

Chamomile, *Chamaemelum nobile*, is best known as a chamomile tea, a simple infusion of flowers made in a tightly lidded teapot. It has proved effective in soothing muscle pain and as a sedative in nervous stomach disorders. After a hustler, Dr. Roger used to administer cups and cups of chamomile tea to a patient who came to him suffering repeated bouts with alcohol. Chamomile tea, together with an injection of B_{12}, quieted the man so that he did not shake; the man always left the office vowing "Never again." For quieting an infant with colic, Dr. Roger recommended chamomile tea for the mother, with a bit in the baby's bottle. The white daisy chamomile florets and short stems are distilled with steam. Volatile properties vaporize in a faint bluish distillate before 180°F (82°C). As the still heats, the vapors become a deeper blue. Most volatile oils are said to complete vaporizing at 300°F (148°C). A British pharmacopoeia lists chamomile as a medically prescribed tonic and sedative.

Cherry, *Prunus virginiana* and *P. serotina*, choke cherry and wild black cherry bark is gathered through the year by chopping and discarding the outer layer and slivering the reddish inner bark into a box. A fragrant odor rises during this operation, but it is gone when the chips dry. The chips should be dried in the shade and stored in paper. A tincture of cherry may be prepared by filling a bottle with fresh or dried, slivered inner cherry bark and filling the bottle with 80 proof vodka. Store in a cool dark place for a month. After the time is up, pour the liquid into a new bottle, lid tightly and use as a cough medicine, one spoonful as needed.

A cough is a reflex action to an irritation anywhere along the respiratory tract or the pleural lining of the lung. Sometimes it is a simple nervous act. Whatever the cause, coughs are controlled by a cough center in the brain. In theory this old cough tincture depressed the nerve reflexes in the cough center. Cherry bark distillates contain hydrocyanic acid, an early-day treatment for whooping cough. Distilled cherry bark water was said to be employed externally for drying skin eruptions as well as internally for bronchial coughs. Cherry bark tea is taken by my mountain lady friend for soothing coughs. The volatile oils of cherry bark vaporize concurrently with boiling water in a still.

Comfrey, *Symphytum officinale*, *S. peregrinum*, common comfrey and Russian comfrey have been called nature's medicine cabinet because comfrey is said to be safe and effective as a remedy for so many complaints. All parts of the plant—leaf, root, stem—fresh or dried, are known for healing. Comfrey is edible. Young comfrey leaves, fresh in salads or steamed like a vegetable, are said to contain about 20% protein. It is one of the few plants that stores vitamin B_{12} and allantoin, a blood-cell builder. Aside from its food value, comfrey greenery or flour milled from the dried leaf and root is said to aid hay fever, asthma and rheumatoid arthritis. Comfrey ointment, made by macerating chopped leaves in warm oil, has a great following among people with bug stings, athlete's foot, burns and sore feet. Comfrey tea has been taken internally as a pleasant drink, and it is reported to have helped people with plantar warts, psoriasis and other skin problems when the parts affected are bathed in a comfrey infusion. As an emollient, used externally, or as a demulcent, taken internally, comfrey infusions are popular to soothe mucous membranes. Stillrooms of yore depended upon comfrey leaves as a poultice to slow bleeding and reduce swelling.

One of the most important constituents of comfrey root is its abundance of mucilage. The young lady who introduced me to my own Pandora's Box comfrey plants explained that comfrey roots are bruised and squeezed of their juice. According to my friend, the old Scottish remedy for stomach ulcers was to drink a large glass of water followed by four teaspoons (20 ml) of fresh comfrey juice, three times daily. If I had an ulcer I would give comfrey a try; it is slippery and should coat your insides from stem to stern. I have not distilled this plant, for I am afraid that it would glue my apparatus together.

Elder, *Sambucus canadensis*. If comfrey has a competitor in the medicinal plant world, it is elder. But elder has been associated with magic, too. A neighboring electrical engineer, a mod-man who drives an ultra auto and skinny-dips in his marble pool at midnight, will not cut his volunteer elder bush lest the Earth Devil invade his life. An old aunt never burned an elder reed lest it "burn the hands of hanging Jesus." If a baby, given a dried elder stick, sucked on it, he would live a life of peace; if the infant pushed the stalk away, anxieties would rule his life. This belief was told to me in earnest by a young couple who proudly announced that their son had sucked the elder rod. (I shivered when they said that; the last elder stalk I looked into exposed a big eight-legger looking back at me.) A remote couple welcomed elder growing along their bottom pasture because they said "witches live in elder bushes and will not wander far." A Missouri farmer, Mr. George, said that elder branches put into the loft would protect his barn from evil, and he swore that cattle warts would disappear if rubbed with green elder leaves. To sleep on a pillow made of dried elder blossoms is to invoke wondrous dreams, a city naturalist said.

Elder flowers have been used in medicines fresh, as a tea for "women's complaints," and infused in wine as a heart stimulant. Dried elder flowers were rubbed into a powder, mixed with oil, heated in a double boiler with a bit of paraffin, and rubbed onto rashes or poison ivy after the area had been cleansed with soap, water and alcohol. Distilled elder flower water was used as a lice deterrent, a sciatica rub, a hot poultice and a corn remover.

Elder always reminds me of the Edwins, our neighbors when I was a girl. It was easy to see that Jim Edwins was suffering "the slavery of youth." He wanted to fly; to him heaven was turning 18 when he would graduate and join the Army Air Corps. Between Jim and heaven stood his father, a man who was as hard as the rocks with which he worked. Mr. Edwin was a stonemason in Missouri who built formidable walls and believed that if God meant man to fly, He would not have put so much effort into building man's biceps. I knew the Edwins and their impasse. The father wanted Jim to follow his steps in the stone trade; Jim was adamant about joining the Corps. The father would not sign the papers and Jim would no longer mix mortar, as he had done since he could handle a hoe.

Then one day I saw the two of them in Mrs. Waldron's backyard; they were building a flower still. That was fortunate for me because when I stopped to say "Hi," Mrs. Waldron asked if I would pick elder blooms for her at 5 cents a bag, and without a moment's delay I leaped into the commercial arena. Daily I saw the Edwins and learned that because of the urgency to get the still built before flowers had waned, Mr. Edwin had given Jim permission to fly if he would help out on the Waldron job. Mr. Edwin was sweet on widow Waldron.

Bricks flew into place, creating an apparatus different than any I have seen. The still consisted of three graduated cast-iron lard kettles that fitted into a brick flume, the lower end of which was a firebox. Sitting on the rim of each iron kettle was an oak barrel with a lattice bottom. The condensers were inverted conical metal lids to the barrel vats. Each of the three boiler kettles was filled with water, the wooden vats were placed on top, and mesh cheesecloth bags of fresh or wilted elder flowers were lowered to settle on the perforated vat bottoms. A fire was lit in the fireplace, heat and smoke emerged from the side of the firebox, the kettles were heated and the smoke exited through a tall chimney at the top of the ascending brick flume.

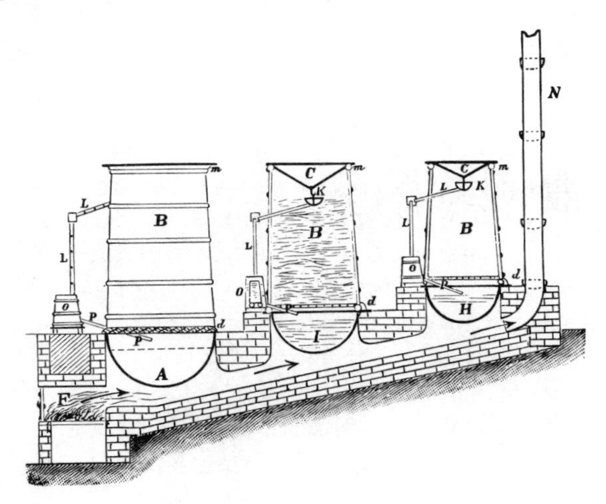

INVERTED CONE STILL

Under the drip point, the bottom of each concave vat lid, a collector cup was attached; it had a downward slanted exit tube that emerged from the barrel. Vapors from the boiling water below each barrel swirled through the blooms, rose and hit the inner surface of the inverted cone lids. The lids were kept filled with cold water, and when the water and flower oil vapors hit the cold lids' inner surfaces, the steam condensed, ran down the lid into the collector, thence through the tube connected to an exterior receiver.

This Waldron still was the inspiration for my Bird's Nest Cove water distillation device. I do not know what Mrs. Waldron did with her distillates, but in addition to cosmetic uses, medieval distillers prescribed a spoonful of Aqua Sambuci (which is an elder flower distillate) in hot water as a tea to remedy sore throats and bronchial afflictions.

Epilogue: Mr. Edwin married the widow Waldron and Grandma told me that he was as "proud as a prince" when Jim earned his wings with the Army Air Corps. "It was as if the boy were carrying on the torch his father had given him," she said with a chuckle.

Eucalyptus, *Eucalyptus globulus*, trees, which grow in the same climatic conditions as orange trees, produce leaves that are high in essential oils and are considered strong antiseptics. They have an unpleasant odor that produces coughing, so care should be taken when handling eucalyptus distillates.

An old fashioned preparation called eucalyptus lotion was used externally as a scalp conditioner, and for callouses, swollen neck glands and joint pain. Fresh or dried eucalyptus leaves were bruised and laid loosely on a rack in a pot still, covered with water and gently heated to around 200°F (93°C). Eucalyptol passes over at about that temperature or a little before. Distilled eucalyptus water is said to increase cardiac action. Cork distillates tightly and store in the icebox. A thick, yellowish eucalyptus oil, may form on the top of the distillate. Herbalists recommend dilution before use. Eucalyptus lotion is prepared by heating 1½ cups (375 ml) of elder flower water to body temperature and stirring in four teaspoons (20 ml) of distilled and shaken eucalyptus water. Bottle, lid tightly, and shake the lotion before applying externally to your callouses, scalp, neck or arthritic joints. Rub in well to increase circulation.

Eucalyptus water has been used as an inhalant for asthma, diphtheria decongestant and sore throats. Excessive inhalation can act as an irritant and arrest respiration or cause coughing. Extreme care should be taken when utilizing the oils of this plant, the literature warns.

Garlic, *Allium, sativum.* Garlic oils obtained by distillation in a noncopper still under diminished pressure contain disulphide, the bearer of the garlic odor. The active properties are rich in sulphur, and if pure, they contain no oxygen. Garlic oil is so strong that if it is rubbed on the soles of the feet, the odor is said to be smelled on the breath. "Who wants garlicky feet?" Lewis asked when I passed on that bit of intelligence. Years ago superstitions held that if garlic were chewed by a man running a race, the odor would prevent other racers from beating him; even today some horse racers give their steeds a garlic bulb in the belief that it will help him outrun his competitors. Garlic has been around since antiquity, and old books indicate that a decoction of garlic water was prescribed for lung troubles. Mixed with lard, garlic was used as a rub for chest congestion; sniffed, it was said to relieve hysteria, and rubbed on the pate, it was believed to be a stimulating lotion for baldness.

A wash made from garlic bulbs crushed and infused in vinegar was concocted in the early 1700s as protection against the plague. The pungent infusion was called Vinegar of Thieves and was said to be used by priests, physicians and robbers as they visited penitents, patients and corpses to be plundered. The odoriferous but immune men chewed garlic and doused themselves with the garlicky vinegar before roaming the plague-ridden streets, and they survived.

It is known today that the plague, which is very much alive and thriving in numerous parts of our planet, is caused by the bacterium *Pasteurella pestis*, which is generally carried by the rat flea *Xenopsylla cheopis*. Plague is essentially a disease of rodents. Fleas acquire the bacilli by feeding on a plague-infected rat, and the bacilli multiply in the fleas' stomachs. The rat dies, the infected fleas find another host—dog, wild creature or man—and the plague bacilli infection is introduced with a flea bite.

Plague illness, called bubonic after the bubo, or pustule, believed to indicate the point of inoculation, develops rapidly. High fever, headache, prostration and delirium often appear within two days after the flea bite. The characteristic bubo contains turbid fluid teeming with plague bacilli and is most often found in the thigh region on humans. Initial symptoms of plague infection include tender glands, skin lesions and faint red lines over the lymph nodes. Blood, liver, spleen and pulmonary involvement follows, and in about a week, if the patient survives, the bubo pustule breaks down, leaving a slow-healing sore. The pulmonary forms may be transmitted between humans by respiratory secretions.

The plague fleas and their bacteria still exist among rodents. Sporadic outbreaks occur throughout the world; with intercontinental travel so prevalent today, a plague epidemic of monstrous proportions smolders, and there does not seem to be a world-wide rat-flea wipe-out plan. Several preventive measures such as immunization, plus sulfa, streptomycin and other drugs are useful to reduce plague dangers, but it might be advantageous to remember the old vinegar-garlic rub and garlic ingestion custom.

Garlic has been well-known as an antiseptic; the ancients considered it a healing deity, and as late as Chaucer's day, garlic was thought to be a heal-all. Some modern nutritionally oriented men of medicine recommend garlic, eaten raw daily, for a healthy heart. Among northland people, wild garlic juice was expressed into sphagnum moss and bound onto open sores.

People of my grandmother's time also held great faith in garlic. When I was sick with so-called rheumatic fever, a neighbor came over and rubbed my chest and "glands" with lard mixed with bruised garlic cloves. I smelled like spaghetti sauce. (I also had to take a spoon of baking soda in water after each meal, then hold a cold silver knife to the base of my throat to keep from throwing up. The cold knife trick really worked.) The neighbor who rubbed me with garlic said she took crushed garlic in honey for her rheumatism and that she gave her husband garlic mashed in a spoon of water every morning for his heart condition. I later discovered that he followed his heart medicine with a shot of White Mule. He certainly was peppy. He trimmed the top off their sixty-foot sycamore when he was 73 and nearly gave the neighborhood several heart attacks. In addition, he hated cats and regularly ran them out of the barn loft through the upper hay door, leaping after them as they sailed to the manure and straw pile below.

The most universal use of garlic is as a vermicide, a worm destroyer. Two tablespoons (30 ml) of minced garlic, mashed with, then strained from ¼ cup (60 ml) of cold milk, nose held tightly and swallowed quickly, does the trick. For free loading dog worms, a spoonful poured between a pup's eager chops while he is begging for a meat ball held in the opposite hand, and he will not know what hit him. Give him the meatball as a chaser.

Ground Ivy, *Glechoma hederacea*, Gill-over-the-ground. Gill water may be made by pulling handsful of the whole stinking herb when it begins to bloom, snipping off and discarding roots, rinsing the plants and stuffing them into a copper pot still. Cover with water and boil until one-fourth of the original liquid is recovered. As a tonic or astringent, gill water, which loses its penetrating odor in the still, was recommended in the amount of one-half teaspoon (2.5 ml) in a wineglass of wine or water, plus a little honey. Combined with chamomile, gill water was used as a poultice for skin eruptions. Dr. Roger made up ear drops by mixing rubbing alcohol with a little concentrated gill water. He advised patients suffering from ringing to put three or four drops of the alcohol into each ear every day. I tried it; the alcohol tickles and feels good but I cannot vouch for other peoples' head noises.

Hops, *Humulus lupulus.* Freshly dried female flowers, the leafy conelike strobiles that appear to be sprinkled with yellow powder, are put into a double cheesecloth sack for distillation. Hops that have been bleached with sulphur should not be distilled in copper. Cover the hops with water and boil vigorously until one-fourth of the original water is distilled.

Although some vapors come over when water begins to boil, higher temperatures seem to be necessary to volatilize the major constituents of hops. One early-day chemist wrote that after boiling the still vigorously, he distilled one-half of the original liquid, covered the distillate and allowed it to settle for at least an hour before pouring off the top fourth. Mixed half and half with 180 proof ethyl alcohol, this was called Tincture of Hops. During stillroom days a nip of this tincture, which contains the bitter, bitter lupulin, was taken for sleeplessness. The lower portion of the hops distillate was mixed with the water remaining in the boiler, allowed to settle and strained into malted barley for use in making beer. If you buy hops, the strobiles should be greenish. Store in the freezer. Old hops flowers may be identified by their dark color and off odor, which indicates that the lupulin has been dissipated; they are no good. Dr. Roger prescribed hops water taken in water or wine as a tonic to improve the appetite and promote sleep. Some people used to cut homemade wines with the hops water for a nervous stomach, and society dames infused hops in sherry as a cordial. To me that seems a sure-fire way to kill a party. In recent literature I have read that physicians prescribe a concentrated hops distillate, sans alcohol, for the DT's.

An infusion of one tablespoon (15 ml) of hops strobiles and leaves in two cups of boiling water creates a bitter tea, which at one time was regularly taken as a spring tonic. A lady for whom I worked in northwest Missouri took her tonic during the rush of lambing one spring, and she slept through the whole meleé of bleats, wobbly legs and anxious mammas in the lambing shed. When she awoke we had 18 babies; luckily the ewes knew what to do.

Horehound, *Marrubium vulgare.* Fresh leaves and flowering tops are laid loosely on a rack in a copper pot still, covered with water, and the still is brought almost to a boil. Simmer at 210°F (98°C) until one-fourth of the original liquid is condensed. Horehound water, which contains the acrid marrubium, was prescribed as a medicine for coughs and bronchial disorders. Some herbalists recommended it be taken in a potable spirit called Bull's Blood, which contained distillates of peppermint herb, rhubarb root, mandrake root, aloe, scullcap leaves, sassafras root, cinnamon, and oils of juniper and sweet birch. After maceration in 190 proof ethyl alcohol, the tincture was incorporated in a horehound syrup base at the ratio of 10% and bottled within a container showing a ghastly picture of a bull being bled. Old medicine labels were graphic. I remember one bottle showing a lady in her 1890's undies with two ropes around her waist. Numerous unfriendly dwarf men were pulling on the ropes and squeezing her kidneys. I used to suffer for the lady when I saw the bottle. It must have been a kidney medicine.

Horehound syrup was made by boiling equal parts of distilled horehound water and sugar. Horehound cough drops were made by boiling three pounds (1.5 kg) brown sugar with one-quarter cup (60 ml) distilled horehound water or a strong, strained decoction of fresh horehound leaves, plus one-quarter cup (60 ml) of distilled boneset water, or its decoction. After the sugar was liquefied over a low fire, the flame was raised and the candy boiled briskly until a hard ball was formed when tested in cold water. The candy was poured into a flat, heavily buttered pan, and just before it hardened, it was marked into squares with the point of a buttered knife. It is best not to stir hard candy while it is boiling nor scrape sugar from the pan sides when pouring into the cooling tray.

Knotweed, *Polygonum*. Native peoples around the Bering Strait made an ointment from the little greenish to pink-and-white knotweed flowers that they believed healed rope cuts, burns and all kinds of skin sores. They melted reindeer fat and stirred in as many knotweed flowers as the warm grease would hold. After keeping the mixture liquid with slow heat for a few days, they strained and discarded the *Polygonum*, which the dogs ate. They cooled the grease and carried hunks of the tallow with them on their travels. It was also good for greasing sled runners so that they would not freeze down, one man told me.

Labrador Tea, *Ledum decumbens*, is another northland native remedy. Infused with boiling water, it was drunk as a pleasantly spicy tea to soothe sore throats or relieve coughing. The high tannin content probably acted as an astringent to shrink tissues and probably did help people rest and feel better. Men drank Ledum tea laced with whiskey and pronounced it a healthful tonic; but one wife warned her husband that too much "tea" would give him a hangover.

Licorice, *Glycyrrhiza glabra*. The sweet root of this warm-climate plant has been a favorite additive to cough remedies since earliest times. Cultivated by the Black Friars of Yorkshire, this tall perennial herb with a leaf like an ash has pale purple flowers and a big thick root that sometimes goes down into the soil an arm's length. Although licorice may be extracted with alcohol and distilled, Dr. Roger used to buy the dried, brownish-yellow licorice roots and soak them in water for a couple of days before boiling them in their water, pressing and straining the product. He later reduced the decoction and I stirred so the aromatic syrup would not burn. To the thick, brown licorice decoction he added jelly from soaked linseed and a skim of ethyl alcohol as a preservative. He prescribed the medicine as a demulcent to soothe sore throats. I used to sneak a piece of peeled root to suck on, but I got caught when it turned my teeth orange. Dried licorice root is very sweet and agreeably spicy; the early Greeks noted that it was the only sweet that quenches the thirst.

Linden, *Tilia europeaea*, also called lime tree, grows well over 100 feet tall, and when it flowers, bees move in and out of the yellow blooms with such passionate determination that the whole neighborhood appears to vibrate. Linden honey is regarded as medicinal and is used in cases of chronic indigestion and nervous vomiting. For centuries Europeans have employed lime tree flowers as a curative tea for "vapors" in women and sleeplessness of the aged. In addition to an infusion, lime flowers have been traditionally distilled into Tilia Water. Pick newly-in-bloom flowers because if they are old or wilted, they may produce symptoms of narcotic intoxication. Fill a steam or boiling-water still three-fourths full, add water in the amount of one-fourth the volume of the flowers, fasten the head, and fire with a moderately high heat. The active properties of linden flowers are fairly volatile and come over before water boils. The distillate is fragrant, smelling somewhat like fresh flowers, but it dissipates quickly, so cap tightly and store in a cool, dark place. In France four teaspoons (20 ml) of Tilia Water in a cup of piping hot water was prescribed for young women suffering menstrual cramps. I have read that physicians of old made charcoal of lime wood, which they pulverized and gave in cases of dyspepsia.

A HOME STEAM STILL

Melissa or lemon balm, *Melissa officinalis*, was the backbone of many kitchen herb gardens during stillroom days. Melissa Cup was a warming winter drink made in August and said to make a man frolick. Just as the tidy little plants were beginning to bloom, the leaves were stripped and put on a rack in a pot still. The herb was covered with water and boiled gently until the distillate began to lose fragrance. This melissa water was mixed with freshly expressed lemon juice, added to elderberry wine with a sliver or two of lemon peeling and bottled securely. When needed to cool feverishness caused by flu, it was served warm in cups. Dr. Roger kept a jar of distilled melissa water, commercial oil of lemon and rubbing alcohol in his treatment room for use externally as an antiseptic wash for superficial wounds. The fresh lemon-mint fragrance seemed to reassure patients that their hurt was not severe. Lemon balm begins to vaporize at about 172°F (77°C), with the most pungent properties coming over just before water boils. It is soluble in 140 proof ethyl alcohol, but it clouds the solvent.

Mint, peppermint, *Mentha piperita*. The months I stayed on the sheep farm in northwest Missouri I learned that mint meant more than candy. I had paid little attention to the crop in a neighboring field until one day our whole countryside seemed to float in a purple mist; the mint flowers were beginning to bloom. No sooner had I become aware of the pearly haze than it was gone. Like an army of locusts armed with scythes, men moved in to cut the crop, causing the August air to lie heavy with the exhalations of dying plants. Mint fumes were everywhere, even our cereal tasted like peppermint because the cow's milk had picked up the odor. The mowed mint was allowed to wilt in windrows for several days, and everyone prayed that the rain would hold off. Just as the farmland again began to smell of cows and sheep, men driving teams with hay wagons arrived to fork the mint into heaps and load it for transport to the herb still. The lady for whom I worked said I could go and watch.

The old still boiler recently had been converted to a steam still, and for a while the distillers argued in the shed. Everyone seemed to have a different idea about "getting up steam." It was a community still and loads of mint kept arriving, horses stamped, children raced around, women stood gossiping in the shade until, without warning, a piercing whistle cut the air. "Steam's up!" As several teams bolted across the flatland and kids jumped with excitement, the distillers sprang into action.

Two large wooden vats received bucketsful of whole peppermint plants; steam pipes passed from the boiler under the racks that held the mint; and small jets of steam were directed through the mint from below. As everyone watched, two men jumped into the vats, trampling the herb to bruise it. When they were barely visible through the rising steam, they crawled out and the stills were sealed. With a giant hiss, a full charge of steam passed upward through the mint and the whole area shook. Once again, kids leaped in circles, horses snorted and the women stood back. I nearly wet my pants; it was very exciting.

The steam, carrying the volatile properties of mint, arose in a large copper pipe that connected with horizontal condenser pipes, which were immersed in a narrow canister of running water. The well water used to cool the condenser was pumped by a blindfolded horse going around a treadmill behind the shed. Everything smelled of mint. As the steam condensed in the piping, it trickled into a square, copper-lined collector box with two taps; the lower faucet ran distilled water back into the boiler as a replacement for that which went up in steam. The upper faucet drained the mint oil that floated on top of the water in the collector.

Several days passed before the mint wagons stopped passing en-route to the still, but for weeks the odor of peppermint permeated our lives. Our eggs, gravy, even the bread tasted like chewing gum. My employer had me empty the butter into the soap fat crock. Although a happy-go-lucky fragrance at first, peppermint seems to eat its way into your pores. I am thankful that Lewis is not a peppermint farmer.

The medical virtues attributed to peppermint oil come from its alcohol constituent, menthol, which may be as high as 85% in some distillates. Distilled in boiling water or steam, vapors begin passing over at about 200°F (93°C), with the strongest vapors occurring just as water boils. A hard rolling boil will distill peppermint very quickly, and depending upon the size of your load, an agreeable peppermint water may be distilled in a few minutes.

A spoonful of peppermint water taken plain or with a bit of sugar has traditionally been employed with antacids to allay abdominal pain associated with flatulence. Warm peppermint water held in the mouth temporarily relieves a toothache. In certain kinds of dyspepsia, such as overeating, peppermint water cools and relieves nausea. Given in hot water it is good for inducing perspiration. Fumes from heated peppermint water have been used as an inhalant to alleviate sinus pain. Glycerine was often mixed half and half with peppermint water as a soothing lotion for external application on hemorrhoids. Glycerine-peppermint water used to be listed as a curable in cases of neuralgia and lumbago; rubbed gently at first and later more vigorously, it was said to act as a vascular stimulant, and afterwards as an anesthetic to localized pain. Some people heat peppermint water hot-to-the-touch, briskly wring out a cloth in it, and apply the damp hot cloth to an arthritic joint.

Mint, spearmint, *Mentha spicata*. Though a taste cousin to peppermint, sharing its cooling property, the milder spearmint contains carvone, a fragrant ester that relieves gas and colic. An ester is a combination of an alcohol with an acid, usually associated with the elimination of water. For medicinal uses the herb must be cut just before it is coming into flower and distilled as soon as possible. Heat and moisture after harvest will cause the plant to lose its volatile properties. Spearmint volatilizes at about the same temperatures as peppermint, just a few degrees before boiling water. Commercially, 350 pounds (175 kg) of spearmint is required to produce a pound (500 g) of spearmint oil. Commercial oils are far more concentrated than home-distilled waters, and they should be diluted with water before use. Make certain if you purchase oil of peppermint or spearmint that it is for internal use; essences manufactured for use in fragrances may be toxic.

Dr. Roger prescribed four or five drops of spearmint oil on a sugar cube to relieve gas. He gave spearmint water sweetened with sugar syrup to relieve hiccups in children. They loved the attention and the taste and forgot the hiccups. He mixed spearmint oil with ethyl alcohol and water and had mother dip her finger in it and allow her infant to suck her finger for relief of colic. I do not recall proportions.

Mullein, *Verbascum thapsus*. This wooly distant cousin of the *Labiatae* family yields its fuzz to tinderboxes, has been beaten to form poultices, distilled for complaints of bleeding bowels and smoked for asthma, but my former teacher, Dr. M. L. Fernald, suggested that mullein contained poisonous constituents. I have read that Indians made a salve for burns by soaking mullein flowers in melted grease, recharging the fat with fresh flowers for about a month and applying the bactericide oil to flesh wounds. Some stillroom matrons soaked dried mullein leaves or roots in boiling water and applied them externally as a poultice for toothache, neuralgia or cramps. I have not distilled the plant.

Mustard, *Brassica hirta* or *B. nigra*. Seeds of both white and black mustard contain active principles that were employed as a counter-irritant years ago when external application of irritating poultices were popular. Mustard flour (ground and sieved seeds) mixed with water, vinegar or spirits was considered an antiseptic and deodorizer when used as a footbath. The sloppy footsoak was said to stimulate the flesh and help throw off a cold. Pastes made with mustard flour and water were applied externally near the seat of an inward inflammation, generally in the bronchial area; they were said to draw the blood to the flesh and flush poisons. Mustard plasters, a dryer mix made with mustard flour, water and a few drops of oil of mustard, were applied externally for neuralgia. Commercial oil of mustard, made from expressed seed husks that have been fermented and distilled, will blister the skin, and they, like mustard pastes and plasters, should be used with caution.

Nettle, *Urtica*. In spring, nettle tops may be picked if you wear gloves; the wilted herb is distilled in a copper pot still. Collect one-fourth of the original water. A spoonful of nettle water, mixed with an equal amount of honey and sipped slowly, has been recommended in asthmatic coughing attacks. Nettle water heated with thick cream and taken as an old-timey soup is a remedy for "huckle bone ache," my mountain friend told me. She also said that a "whipping nettle" was good for rheumatism; the ritual, she explained, consisted of a wife "getting her old man to whip her good with nettles." The sting of the plant was thought to enter the body and allay the pain of rheumatism. Preparations made with fresh nettles are said to be useful in tonics because they contain mineral salts. Herbalists make tinctures by macerating the whole herb in ethyl alcohol and administering it for chills and fever.

Oak, *Quercus*. Strip the horny outer bark and discard. Chip small pieces of inner oak tree bark and lay loose on a rack in a boiling water still, cover with water and boil. Do not connect the head. Close down the still and allow the bark to sit in its own decoction overnight. The next day connect the head and coil, and boil the still vigorously, collecting one-fourth of the original liquid as it comes over. Oak bark water has a bitter, astringent taste, but it was extensively used in the Middle Ages to counteract diarrhea. A spoonful in a wineglass of water or wine was recommended. In a farm magazine I read that too many acorns, which contain many of the same properties as oak bark, can cause pig constipation, so I would be careful about taking excessive amounts of the distillate. Oak bark water has been reported to be beneficial as a gargle for sore throat and for sore gums. The ancients believed that the tannin present in a decoction of oak bark and buds deterred the spitting of blood that accompanied consumption. Later observations have reported that, for whatever reason, few tanners contract tuberculosis.

Pine, *Pinus*. Several years ago I saw a man extract resin from peeled, slivered and chipped yellow pine roots. He told me that preparing the wood was hard work but that after the pine had been chipped, distilling pine oil was child's play. First he dug downward from the top of an old road-cut bank to make a hole big enough for him to sit in. Next he placed a 90° elbow sink drainpipe in the hole so that the pipe pointed upward in the center of the hole and out the cut-away bank with its opposite end. Propping it in place and putting a tin can cover over its top, he made a big hot fire in the hole. When he had a deep bed of red-hot coals with the upright pipe end barely showing through the center of the coals, he placed a metal sink on the coals and fitted its center drain opening down into the drainpipe below. The pipe was wider so that the sink outlet easily slipped inside. He pushed the sink into the coals.

Working fast, he filled the sink with pine chips, lidded the sink with metal, shoveled dirt over the coals, sink and the side bank with its outlet pipe. He heaped up dirt all around so that the only thing showing was the end of the drainpipe that emerged from the side bank. Then he had a beer and waited. After a while clear watery fluid trickled out from the pipe. As the flow increased and became yellowish, the man repeatedly tested the effluent. He had me feel the stuff. It was sticky and smelled like turpentine, but I tasted it. It was turpentine. The man put a container under the drip and a surprising amount of the sap poured out from this downward distillation. As it flowed the liquid became thicker and darker. Resin, he pronounced the last turgid, brown string. This Pennsylvania farmer sells his home-distilled turpentine-resin to soap and pharmaceutical manufacturers. He saves a little for liniment, he said, which he makes by mixing the distillate half and half with olive oil. Rubbing it on "keeps a body souple," he told me with a twinkle in his eye as he said "soup." Commercially, the distillation of pine for turpentine and tar is done by means of steam under pressure. Someday when I feel the need for keeping "souple" I plan to hunt up yellow pine roots and distill them by the side of the road.

Plantain, *Plantago major*. Before the plant shoots to seed, cut and bruise the leaves and place them loosely in a net in a copper pot still. Cover with water and boil the still until one-fourth of the original liquid is recovered. As a mouthwash, distilled plantain water has been reported to be antiseptic. Ringworm infections used to be treated with a plantain water wash, as did open wounds and acne. Plantain seeds are considered excellent as a mild laxative and have been substituted for linseed. Fresh, wet plantain leaves, externally applied, were used by stillroom wives to arrest bleeding, soothe burns and insect bites, and reduce hemorrhoidal inflammation.

Potato, *Solanum tuberosum*. Early Americans held the potato in high regard as a remedy for rheumatism. Raw potato juice was warmed and applied with a saturated cloth to the affected area. Sprains were treated similarly by immersing the joint in hot potato water. My grandmother successfully treated my ankle that swelled like a balloon after I became tangled in the curtain ropes as I rushed off stage. I was performing a comedy dance, but that final sprawl was not part of the act. No one knew I was hurt but when Grandma saw my foot she quickly ground lots of potatoes, skins and all, heated the sloppy foment and had me soak for an hour. She kept reheating it. I was dead tired. Finally allowing me to go to bed, she elevated my ankle (and knee) with pillows and by morning the swelling and soreness were gone. Because potato is a member of the nightshade family, the green parts of the potato plant—stalks and berries—are poisonous to eat. But when cooked with their jackets on, potatoes contribute potash and phosphorous, which are valuable in today's salt-skewed diets.

Raspberry, *Rubus idaeus*. What potato fomentations did for lumps and bumps on one's outside, my grandmother believed raspberry did for one's insides. According to her, raspberry vinegar, juice, syrup, leaf tea or wine could cure a girl's anything. Raspberry vinegar was her standby. Fill a jar with raspberries, cover with natural cider vinegar, drape with a cloth to discourage flies, and set the jar in a warm place. For the succeeding three pickings (two or three days apart) pour the vinegar off the infused berries and pour it over a fresh batch of berries. Let the last batch set three days in the vinegar, which gets darker. With a double thickness of cheesecloth laid in a collander over an enamel pan, empty the final vinegar infusion and allow to drain. Using equal parts juice to sugar, stir in the sugar. Over a low fire bring the liquid to just a simmer. Remove and bottle. According to the old ways, a few spoons of raspberry vinegar in a small glass of water made a remedial gargle for sore throats. Ingested straight it tightened the bowels. Raspberry vinegar was said to settle the stomach, allay fever, retard tooth tartar and cure a headache, but I drank it in cold water like pop; I would sneak some anytime I had a chance.

Rose, *Rosa*. The medical virtues of roses rest in their volatile oil found in both the flower petals and in some new leaves. I distill any rose with a fragrance by putting the petals in a net bag in the copper pot still and covering them with water. Or I fill the steam still with petals. I boil the still until one-fourth of the original liquid comes over into the collector. Roses become volatile with boiling water; the distillate is clear and the water in the still is rich red. For color, some distillers concentrate the distillate then dilute it with filtered water from the still. Rose water virtues are legendary; it has been prescribed in cases of jaundice, joint aches, fainting, trembling and weak stomach; it has been considered a tonic, astringent, diaphoretic and stimulant. Not the least rose virtue is the pleasant odor that it lends to pharmaceutical preparations. To me, rose water tastes like a blessedly fragrant, tannic-rich tea.

Rosemary, *Rosmarinus officinalis*. Upper parts of new shoots are taken together with leaves stripped from woody parts; however, flowering tops, are said to contain the best oils. Dried or fresh rosemary is distilled by laying the herb loosely in a copper pot still or in a steam still. Because Rosemary Oil contains five major constituents—borneol, camphor, cineol, pinene and camphene—each volatilizing at a different temperature, rosemary herb vaporizes over an extended spectrum of heat beginning at about 150°F (65°C) and continuing to condense into a strong distillate when the still water boils. Lid the distillate immediately. Rosemary water was used by my Alaskan doctor as a pepper-upper. He believed it cured depression headaches and prescribed a teaspoon (5 ml) in a wineglass of water or wine. Some patients liked it with an equal amount of honey. Taken every half hour until four doses had been ingested, then before meals for four days, Dr. Roger said rosemary water never failed to relieve depression. An herb lady I know chops fresh rosemary into her sherry at the end of the day, saying it keeps her blood circulating for her evening chores. Another woman, a hospital switchboard operator who had lively black eyes and feet "made for dancing" (a rounder, my grandmother would have characterized her) kept a good-size rosemary bush in her small house, 125 wooden steps above the main street of Juneau, Alaska. She regularly sacrificed the plant's new growth to her vodka bottle, infusing the leaves and drinking the liquor in Bloody Marys. It made her merry, she believed. She was fun. But one day she did not make it to the top of her steps; they found her in a snow bank: heart attack. It was the way she would have wanted to go, returning from a dance, her son told me, adding that he had planted her rosemary bush on her grave. "Rosemary for remembrance," a symbol of love and loyalty, he said.

The rosemary decoction remaining in the bottom of Dr. Roger's still after he had taken off the oils was mixed half and half with rubbing alcohol and used as a stimulating liniment. He called it Hungary Water, and it was said to be very effective when gently but firmly rubbed on sore muscles and stiff joints.

Sage, *Salvia officinalis*. Sage leaf distillates, which had been volatilized in the still at around 155°F (68°C), were old-fashioned treatments for cooling fevers, cleansing the blood, allaying biliousness and relaxing the throat. Two tablespoons (30 ml) of sage water, when taken with lemon and honey, was a standard cottage remedy of the 1930s.

Sassafras, *Sassafras albidum*, root bark contains a heavy oil plus resins that upon distillation may form a residue in the bottom of the still. For this reason root chips were kept off the still bottom by use of a perforated rack. The still was sealed and boiled vigorously, cooled overnight and the distillate poured back into the still. The still was reboiled until one-half of the original liquid came over. Some distillers ran the same material three or four times because of the bulk and because sassafras oil vaporizes between 208° and 229°F (97° and 109°C).

Dr. Roger prescribed sassafras water, one-half teaspoon (2 ml) taken in a cup of warm, honey-sweetened water as a midafternoon drink to discourage rheumatism. Some people drank sassy water diluted in tea to induce sweating when they thought they were coming down with a cold. Years ago dozens of home remedies employed sassafras decoctions or distillates, but in the 1960s the Food and Drug Administration reported that safrol, a constituent of sassafras root bark, was carcinogenic to rats, so people are hesitant about using the old treatments. A lady in Virginia said she was not afraid of sassafrass; she drinks a decoction of the root to relieve gas. She said she also made her husband rootbeer toothpicks by slivering the sassafras root bark. She said he put one in his mouth when he felt like a smoke. He subsequently left her, so I never heard whether he quit smoking. In the late 1500s sassafras was shipped to Europe and sold as a miracle cure for syphilis, a treatment that was employed as late as the 1930s for many gleety conditions. One-quarter teaspoon (1 ml) oil of sassafras was dropped on a sugar cube and taken every other day for a month.

Sorrel, *Rumex acetosa* and *R. acetosella*. The leaves of these two plants, garden and sheep's sorrel, are well known for their virtues as a popular scurvy remedy. The leaves were mashed fresh and taken with a little vinegar and sugar, or fresh or dried leaves were given as an infusion. Sorrel leaf tea was taken as a mildly acid drink to refresh a feverish person. Sorrel leaves, mashed with a little water, used to be an accepted mouthwash to stanch gum bleeding. Sorrel loses its medicinal qualities when distilled, so I have not tried it in my still. But I enjoy sorrel leaves as a pleasantly tart nibble as I cultivate the garden.

Strawberrie leaf tea/Quit dysenterrie. An old jump-rope rhyme.

Sumac, smooth and staghorn, *Rhus glabra*, *R. typhina*, are tannin-rich plants. Care should be taken to positively identify sumac before touching it. Nonpoisonous species have red, hairy, acid fruit in dense, terminally compounded clusters that usually stand upright in a pyramid shape. After you have made your sumac-berry lemonade in August, put the acid-exhausted berries into a still together with all of the sumac bark you need. Cover the material with water, seal the head, and distill over a lively flame until you collect one-half of the original liquid. Sumac water, a spoonful in a wineglass of water, is an excellent gargle and astringent for puffy gums or hoarseness. Do not swallow or it may shrivel your insides. Sumac water is also said to be antiseptic and effective for external use with ringworms and as a wash for hard-to-heal sores. Foot fungi are reported to clear after several sumac water soaks followed by rubbing the affected area with dry, powdered sumac bark.

Sunflowers, *Helianthus*. Although the oil of sunflowers is cold-pressed and not distilled, a bag of sunflower seeds were generally kept in rural stillrooms for chills. One-quarter cup (60 ml) of sunflower seeds was boiled in a quart of water, set on the back of the range and allowed to infuse overnight. In the morning the infusion was strained and pressed so that oil would weep into the brownish tea. Honey was added, and the recommended dosage was a wineglass four times a day. Men used to take their medicine with a bite of Holland gin, and women with sherry.

Tansy, *Tanacetum vulgare*. Since antiquity, tansy has been employed as a spring tonic to dispel worms and quiet the kidneys. It distills into a camphor-scented product and passes over at about 205°F (96°C). But I threw away my tansy water when I read that oil of tansy can poison a person, causing a condition not unlike rabies. Commercial houses dissolve the dried herb in 140 proof alcohol and distill in vacuo for oil of tansy to be sold as an anthelmintic (dewormer).

Thistles, *Cnicus benedicta*. Though despised by most farmers and animals, some thistles are considered to be of medicinal value. The whole herb is harvested just before flowering, cut into pieces and laid loosely on a rack in a copper pot still. The herb is covered with water, the still sealed and boiled to produce one-fourth of the original liquid. A spoonful of thistle water was generally taken in a wineglass of water as a blood tonic and was said to aid the memory. Though large doses acted as a strong emetic, a small wineglass of thistle water was said to be beneficial as a treatment for tension headaches and rheumatism.

A tea made by infusing a tablespoon (15 ml) dried or minced fresh thistle leaves in a cup of boiling water was considered to have great power in producing milk for nursing mothers. "Thistle tea makes milk," my Virginia mountain friend told me when I asked what she was sipping. I nodded in agreement; she possessed the equipment to provide lots of infant nourishment and she was a beautiful mother. Years ago thistle juice, extracted by mashing all of the crushed leaves that a cup would hold, then filling the cup with wine, was employed externally to cleanse and heal festering sores. The bitter crystalline, cnicin or salicin, soluble in alcohol, was said to be effective in diminishing discharges caused by ulceration of a wound. Young thistle leaves, eaten in salads like watercress, were believed to purify the blood. My donkey browses on young thistle leaves, and I can attest that she has a pure heart.

Wintergreen, *Gaultheria procumbens*, a low shrubby evergreen plant that thrives in cool, damp woods, has provided flavorful tonics, diuretics and astringents for man since he wandered in the northern wilderness and spotted the aromatic bright red teaberry, as explorers called the plant. The volatile oil of wintergreen, which contains quantities of methyl salicylate, is prepared by steeping chopped wintergreen leaves a day or two in water. A chemical change takes place when the bruised leaves are allowed to stand in water. After a few days the leaves are transferred to a boiling water still, covered with fresh water and heated to 120°F (49°C). The still is kept at that temperature for 24 hours to complete the chemical change. After that time is up, the still is vigorously fired. Volatile oils begin to rise almost immediately, with most active properties coming over into the condenser between 160° and 235°F (71° and 112°C). Wintergreen water contains properties of the salicylates, which are antiseptic, though somewhat caustic, and are considered to be drying when used externally on abrasions. Diluted and taken internally, wintergreen water has been reported to reduce fever, relieve pain in muscles and give relief in rheumatic complaints. One old book stated, "although wintergreen decoctions do not cure, they certainly afford relief of the distressing symptoms when rubbed on joint or muscular complaints." The book recommended wintergreen water mixed half and half with glycerine for the rub.

Witch Hazel, *Hamamelis virginiana*. The wattled woods along the Hazel River in Virginia zing with exploding seed pods in June, and when the miniature artillery fire dies down, the season is ripe to gather hazel leaves. The fresh leaves are layered loosely on a rack in a copper pot still and covered with water. In the case of a steam still, the vessel is filled with leaves and a little water is added. In either situation the still is boiled vigorously, and when one-fourth of the original liquid condenses, the still is shut down. Some people run their distillate a second time to concentrate it. I do not, but I save both the residual liquid from the boiler, which is brown, and that which comes over into the collector. The boiler water is used for soaking the feet or for fertilizer for acid-loving plants. The clear distilled witch hazel water, called Liquor Hamamelidis, is employed as an astringent and antiseptic in rubs and ointments. It also is reported to contain sedative and pain-killing qualities when rubbed on a hurt.

My grandmother kept home-distilled witch hazel water on the shelf for chigger bites, stings and skinned elbows. When she daubed it on, the sting went away. When I lost my front teeth I had to wash out my mouth with witch hazel water. It seemed to draw my mouth into a pucker, but the bleeding stopped. Witch hazel water has been prescribed externally for cleaning and reducing the swelling of hemorrhoids. Dr. Roger applied hazel lotion, which he made by mixing Liquor Hamamelidis half and half with glycerine, to inflamed and swollen tissue. Patients said his hazel was a magic lotion because it made the pain go away.

Years before I went to Alaska I heard of home-made tincture of witch hazel. Bruised hazel leaves were macerated in rubbing alcohol for four days before being strained, the liquid was applied externally in cases of inflamed varicose veins. I asked Dr. Roger about it and he concurred that it was an excellent treatment for varicosity problems. Legs elevated, a cloth was soaked in the tincture and laid on the affected parts. The veins were said to shrink. Mixed half and half with pimento water (allspice), witch hazel water is reported to be valuable in check bleeding. The mixture also creates a styptic after-shave lotion.

Wormwood, *Artemisia absinthium.* Artemis was a mythical Greek goddess of the hunter's moon who gave her name to this bitter tonic because it helped her. She must have had worms. Wormwood is an anthelmintic (dewormer), a fact to be considered when ordering the French liqueur absinthe. You can expel your intestinal worms while enjoying an after-dinner drink.

The one time that I was given wormwood I was near-frozen, seasick, and almost a cadaver aboard the banana boat Columbia, enroute to Alaska in August, 1945. Being the first steamer north after V.J. Day, the wallowing tramp from the tropics showed no compassion for the afflicted; no running lights were allowed, nor radio, nor stateroom light because our nook aboard that creaking steamer had no door. My cabin mate and I were the only women passengers, and men peeked over, under and, covertly, through our swinging, saloon-type louvred door as they passed. We did not dare bathe. Actually, I had no inclination to do anything the first day except rush to the bathroom, crawl back into my damp, icy bed and watch a frozen spider rock side to side in his web above me. I had the upper bunk.

As dusk lowered visibility to an occasional line of sea froth leering at us from the blackness outside our solitary window, my shy cabin mate in the bunk below and I lay talking in the dark. She was a dentist's wife going north to be with her spouse; I was heading for the Kuskoquim where I had a job to teach native children. The steamer groaned, footsteps echoed, the boilers belched and always the sickening indulations. Gradually we both dozed.

Whether my awakening was caused by a sound or a wave of nausea, I suddenly became aware that someone was standing inside our room. That person made no movement; he stood silent as a shadow; nevertheless, I sensed his presence. "What do you want? Get out!" No movement came from him because no voice came out of me. Abruptly I knew that I was going to be sick. Swinging my feet, I slid over the side of my bunk and immediately staggered into a cold, wet rain slicker. The man grabbed me and I threw up.

The shame I felt I will never understand. Our cat is not ashamed when she spits up on the porch. Oh the agony of civilization!

In any event, the man left without a word and, shaking uncontrollably, I reeled into the bath. Later I called out to my cabin mate who was petrified with fright. She had seen our visitor outlined in the door louvres but said she could not move. I insisted that she climb topside with me. There we lay like slabs of wood the remainder of the night.

At dawn, one glimpse of the green sea sent me flying to the bathroom again, and when I dragged back to the stateroom I ran smack into a boy of about 15, curly-haired, rosy-cheeked, with a short white jacket, who stood holding two steaming mugs. "From the galley, bitters, good in cases of the flops," he said to me with a wise wag of his head. He handed coffee to my cabin mate.

The dark green, biliously bitter, strongly acrid liquid was wormwood. I never want to taste it again, but within the hour my seasickness left me. Within two hours my cabin mate and I were on the deck drinking in the boundless splendor of that dramatic land. Later we complained to the purser.

A mistake. Apologies, apologies, we would be given blankets, a door, a charcoal heater, fruit. . . . A mistake. We were supposed to be in a different cabin. . . . Men on watch sometimes sneaked into unoccupied cabins for a midnight nap. A mistake. . . . forgive our terrible error. They will be reprimanded, you will be guarded. A mistake, please forgive.

Seven days, seven breathless days aboard the Columbia. "Joy rises out of the bitterness of suffering." Out of the bitterness of wormwood, I would amend the scholar's words.

CHAPTER

Spiritus Vin

While the bonnet holds and the old tube burps,
While the boiler sings and the peeper spurts,
Women and song take second place.
The still is our sun and the moon ferments fate.

KIPLINGING IN THE STILLROOM

Ira McC slid his squat seiner, Hell's Kettle, up to the wharf as he shouted, "Grab a line," and simultaneously heaved a rope toward loafing Alaskans. Reversing, the bow line tightened and hardly had the stern been made fast before the dockside men scrambled aboard. "Hey, not so fast," I heard McC call; all the while the grin on his long face made him look like a fiddle with teeth. "There's plenty of good stuff, all two-year-old, smooth as a baby's behind. Three bucks a pint." He stuffed bills into a chamber pot as he handed out bottles of moonshine.

Though some men brag about making moon, and I have suspected others, Ira McC, known as Spunkie, was the only true-blue bootlegger I have known. Actually, I met him in church. "Hog farmer," our mutual acquaintance introduced him. Spunkie had pigs, a garbage truck, a licensed salmon seiner and an unlicensed still. Labor is known as a poor man's pride, and Ira made what he called a proud drink. He was a distiller of the old school, a man who made whiskey with grain, malt, sugar, yeast and the virginal waters from the island on which his copper still was located. After church sometimes I went with other townspeople to visit Spunkie's island; his was a wide-open operation. He bought sugar, crushed grain and corn meal by the bag, took delivery at the dock and overtly transported spent mash via the Hell's Kettle to his dockside truck, thence to his pigs. "Why don't you keep your pigs on the island?" someone asked. "No Sir, that ain't no life for a young pig. Men and pigs, they learn by imitation, first thing I know my pigs would be making the stuff."

Spunkie McC started his Seward enterprise as a trash collector for Fort Raymond, the Army post established in the 1940s to protect the ocean railhead after the Japanese had invaded Alaska. He fed edible garbage to his pigs, and from that business his whiskey making developed. "There's no better pig food than Army scraps and spent mash," he would assert.

Ira was about 20 years my senior, unmarried, and with money enough to buy a diamond large enough to choke a blue jay. He used his diamond to cut a slash in every bottle of hooch he sold. Few knew of his crescent moon sign. He believed the growing moon to be a lucky time to ferment things. He told me so when he offered me his ring and the job of "permanent slasher." I had had some weird proposals in my years of happy spinsterhood, but to spend life with a moonshiner marking his bottles was not my idea of a rose-covered cottage. He was a nice man; thoughtful, tall, clean-minded and fun to be with, even though he was engaged in making and selling illicit whiskey.

Spunkie aged his liquor aboard a beached and partly submerged barge, which also served as his island dock. Water washed over the land side of the half-buried derelict, but offshore the bow pointed skyward. In the darkness of that canted bow cavern, he stored his barrels. Not many people knew of his hiding place, which could be entered on foot only at extreme low tide. Spunkie McC never sold raw distillate; it was a point of honor that his liquor had been aged in charred oak.

Making Moon the Old Way

His father had been a moonshiner in the east, Ira told me. After prohibition became a law in 1920, his family had moved to Oregon because "racketeers and ornery neighbors" were worse than government men when it came to putting pressure on "honest moonshiners," he said. In the early years his father made whiskey out of a mixture of ground corn and rye. He scalded the grain, cooled it to "body heat" and stirred in malted barley. Later he added yeast, added more water and covered his fermenting mash with quilts to keep it warm. When the yeasts had turned the grain sugars into alcohol, he distilled it.

The process of making spirits is simple. Fermentation is the result of yeasts and enzymes that convert sugar to alcohol. Yeasts, which require oxygen for growth, eat dissolved sugars and with the aid of enzymes produce carbon dioxide and alcohol. The sugars generally come from plants; plants are made up of carbohydrates, proteins and water. The carbohydrates are sometimes regular sugars, as in sugar cane, or they are starches, as in grains or potatoes. Years ago the cheapest sugar available to most farmers was grain, but those sugars were locked up in starches and had to be changed to make them into sugary yeast food.

The Scots and Irish of yore, and before them Asian peoples, discovered that if grain is allowed to sprout it breaks itself down into a kind of sugar with which to feed its newborn seedlings. This breakdown of starches into sugars is called enzyme hydrolysis, or malting, in the liquor industry. Malted grain not only creates sugars from its own starches, but if some malted grain (often malted barley) is mixed with water and other ground-up grains, the enzymes from the malted grain infiltrate and convert the unmalted, new grain starches to simple sugars, which yeasts can eat.

Thus it was that the senior Ira McC made whiskey without buying sugar. He malted some barley—that is he sprouted, dried and ground the barley—and added it to his ground and scalded corn and rye mash. Scalding often means cooking and cooling the grain, or it can mean covering the ground grain with boiling water or boiling residues from the still. In any case, scalding breaks down some of the starches and makes them receptive to the enzymes of the malt. After adding the sprouted barley malt to his lukewarm scalded grain, Ira's father let it stand awhile before he charged his sugar-converted grain mash with yeast and fermented it for a week or so. During fermentation the yeasts fed on the grain sugars, grew and multiplied like crazy, all the while giving off carbon dioxide gas and ethyl alcohol. Then he drained the liquid and fired his still.

Although Spunkie fermented malted grain, he also added sugar to his mash. He said that he copied his father's methods except for the addition of sugar and the layering of bran atop his fermenting mash.

He said on cold nights bears kept stealing his blankets and he did not want his mash to be chilled. I asked him how he kept the bears out of his sweet beer, as working mash is called. He laughed and said that bears hate the odor of urine so he had designed a natural defense system.

With the evolvement of prohibition, Spunkie told me that the demand for moonshine skyrocketed. Instead of the old malted-grain-conversion-to-sugar techniques, which took a lengthy time to be ready for distillation, many illicit prohibition whiskey makers fermented mostly sugars; a little grain was thrown in, yeast and water added, and in a few days it was distilled. A man could make a quick buck using mostly sugar to create "popskull," Spunk said. He added that the druggist in town handled rotgut, that the man had started peddling the cheap booze when the army men were stationed nearby, and that the druggist's shellac had been responsible for some brain-busting hangovers around town. It was apparent that Ira McC had no use for men who bootlegged whiskey made from sugar and a kicker (a kicker is a chemical similar to urea, which speeds fermentation).

I remember that my people, in the 30s, perked off a little something for the medicine cabinet; they used sugar, malt and grain. I told Spunkie so, adding that I never heard of headaches from Uncle Frank's medicine. He agreed, saying that sugar was no culprit when used in combination with malt, adequate grain, fresh yeast and proper aging.

In the end it was the fresh yeast that sent poor Spunk up the river. I had moved to Sitka and had been working there for two years when I heard the story. Ira had found a wife, a newcomer, who sat in the drugstore and drank pop by the hour.

In time she and the druggist were making goo-goo eyes at each other, and before a year passed Mrs. McC was cheating on her husband. Spunkie ordered his fresh yeast by air from Anchorage, and it was often delivered in a parcel with the druggist's refrigerated serums. One afternoon he strode into the drugstore to pick up his yeast, saw his beloved in another's arms and started throwing bottles. The more vials he threw, the madder he got, until, as a witness related, Spunkie was roaring mad and throwing everything his hands touched. The storekeeper and Mrs. McC cowered; when they raised up, Ira beaned them with a bedpan. His tantrum lasted, the story goes, until there was not a bottle left on the shelves.

I had never known Ira McC to drink whiskey—no one had—but that night when they found him in his cabin he was sobbing uncontrollably, and someone gave him whiskey. I was surprised to learn that he had been an alcoholic; that night he suffered a crying jag, and the more he drank the harder he cried. People liked him and tried to help, but his heart was broken, they said.

The druggist put pressure on the authorities. Ira McC was convicted, not for destroying property, but for the sale of illicit alcohol. I have wondered if Mrs. McC took the job of permanent slasher when she married, and what happened to the diamond that would choke a jay, and if the barge with its cache of amber gold still thrust its bow skyward from the North Pacific. I have not been back to inquire.

Whiskeys

Whiskey is a grain spirit made from either corn, rye or wheat, plus small amounts of ground oats and barley, or a combination of the major grains, or from hog feed. Scotch whiskey is distilled from barley, rye and other cereals. The barley is sprouted in peat-infused water and dried; the sprouts are removed, the grain ground, and the barley malt roasted on porous trays placed above peat fires. The roasting of the malt over peat fires gives Scotch and Irish whiskeys their distinct flavor. Some Scotch whiskeys are blended with other spirits after distillation. Irish whiskeys are made from ground wheat, oats, rye, unmalted barley and malted barley smoked over peat fires. Irish whiskeys are generally distilled in copper pot stills.

Bourbon is an American whiskey made primarily from fermented corn that has been distilled in column stills to 95% alcohol and aged in white oak barrels, which give it an amber color. Bourbon that has been cut, aged and sold straight from the barrel is known as straight whiskey. Blended bourbons are made from different straight whiskeys or ethyl alcohol with coloring and flavoring added.

Rye whiskey, often made in Canada, is in large part fermented from rye with lesser amounts of corn, wheat and barley and generally distilled to 80% alcohol.

Most commercial whiskeys are aged and chemical changes that will affect the character occur during the process. In addition, the flavor, color and odor of distilled beverages will be influenced by the presence of volatile oils and other substances present in the fermented mash.

In general, commercial spirits are made by first converting barley into malt enzymes. That is done by soaking the barley in water at

59°F (15°C), draining it and allowing it to germinate for three or four days. Sprouting activates the enzymes that are native to the grain. The sprouted grain is called malt. The malted grain is then dried, the sprouts are removed, and the grain is ground sand-fine. The malted barley is next mixed with plain ground grain. The enzymes in the malt will quickly change grain starches to sugar. Very few enzymes are required to change large amounts of grain to sugar. A bacteria is sometimes added at this time, and the mash is allowed to ferment at a cool temperature so that it will sour slightly. After souring, the mash is heated rapidly to 167°F (75°C) to kill the souring microbes. It is then cooled to 77°F (25°C) for inoculation with yeast of the *S. cerevisiae* family. The mash is slowly fermented at between 59° and 77°F (15° and 25°C) until the sugars have been turned into alcohol. After fermentation is complete, the mash is strained, settled and distilled in steam heated column stills. Condensing temperatures are controlled in these stainless steel, chimneylike stills. The still beer, is heated in the bottom of the still; or sometimes it is heated separately and piped into the still. In either case the rising beer vapors hit a series of cool baffles that are like ladder rungs on the inside of the chimney. Some alcohol and water condense as the vapors hit the cool baffles. Inside temperatures are kept above 173°F (78°C) but below the vaporization point of water so that the alcohol will continually revaporize and continue upward while water drops to the bottom of the still. Alcohol vapors finally reach the top of the still, where they—together with some volatile oils that flavor naturally fermented whiskey—are condensed in a special chamber. From there the alcohol is piped to a special rectifier, which redistills the alcohol and recovers a specified proof liquid. Although some distilleries employ the big-bellied copper kettles in use since the 1800s, most commercial stills are huge, sleek contraptions that automatically measure everything to a gnat's eyebrow in order that the product will be standardized. There is no romance of chance in a commercial still; the operation is overseen by skilled, impersonal technicians and watched over by that great big federal eyeball in the east.

Throughout the United States there are a few unique distilleries that hold on to methods passed down by their founders. The "Old Lincoln County Process" of charcoal leaching corn whiskey is still practiced by the Jack Daniel's distillery in Tennessee.

Tennessee whiskey is made by first cooking ground corn and cooling it. This forms a mush that can be easily infiltrated by malted barley enzymes. Rye and water are stirred into the mash, and when it is cooled to about 146°F (63°C), the barley malt is added. As soon as the malt enzymes have converted the grains into sugar, the slurry is piped to a fermentation vat, charged with yeast and residues from a previous batch, and the ferment is kept cozy while it works. The fermenting brew boils like a simmering soup.

When most of the sugars have been fermented into alcohol, the still beer is separated from its sediments, settled and piped to the tall still. As vapors travel to the top of the still, they are continually redistilled, and finally pass into a water-cooled condenser.

The liquid that dribbles into the collector at the end of the copper condensing pipe is raw whiskey. It is ethyl alcohol plus fusel oil, acids, esters, aldehydes, ketones, volatile oils and tannin.

Elevation.

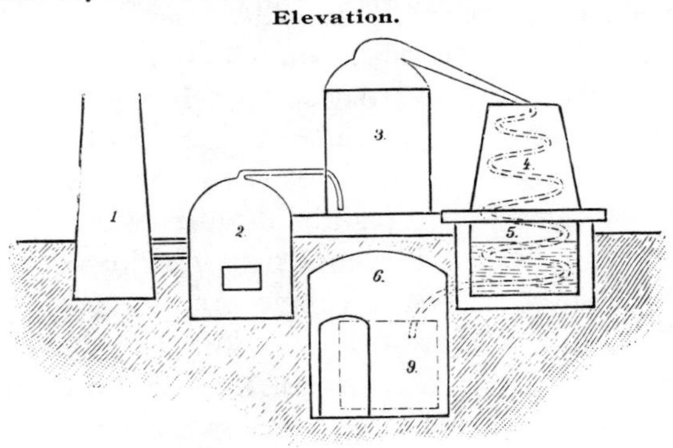

Fig. 60.

1. Chimney. 2. Boiler. 3. Stills. 4. Condenser. 5. Condensing basin. 6. Locked cellar for storage of the distillate. 7. Roof (fig. 61). 8. Water supply pipe (fig. 61). 9. Basin for storage of distillate.

A DISTILLERY LAYOUT

The Tennessee whiskey process involves filtering raw whiskey through ten-foot tall vats of crumbled hardwood charcoal. Natural charcoal is a miracle substance. Because neither air nor water affects charcoal and its porosity creates a hundred surface feet to the cubic inch, charcoal can absorb vast quantities of matter.

Charcoal filtration in the rain barrel used to be commonplace. In Missouri we had a cistern that collected roof water for washing. One time my girlfriend and I heard horrible gargling coming from our under-the-porch cistern storage tank. Lifting the trap door we were astonished to see slit eyes and a set of fangs bobbing amid the charcoal on the black water surface. I thought it was a monster and was ready to drop the door and run; but my friend lowered a rake and thrust it into my hand while she prodded the creature toward me. Suddenly the maddest tom cat I have ever seen clawed its way up the wood, up my arm, dug itself into my shoulder and snarled cat cuss words. Charcoal plastered its fur so that it looked like a prehistoric lizard. My girlfriend doubled over laughing while I tried to get the thing off my back.

Today many people discount the natural qualities of charcoal as a filtering agent, but distillers understand its value. In whiskey making, charcoal filtration is a mellowing process. It filters out many of the harsh congeners present in raw alcohol. From the massive charcoal filters Tennessee whiskey runs into United States government vats, where it is barreled, gauged, recorded and stored in bonded warehouses for maturation.

Most whiskeys are matured in white oak barrels. In the United States, whiskey is diluted with water (sometimes special "spring" water or distilled water) and barreled at 103 proof: 51.5% alcohol. Scotch whiskey is matured at 124 proof; brandies are stored at 140 proof. When spirits are matured in wood, air slowly penetrates the wood; if proper temperatures and humidity are held, some evaporation of water in the spirit occurs through the wood. The alcohol and volatile oil content of the beverage get stronger. As concentration occurs, the component parts of the whiskey, the acids (some of which are extracted from the wooden barrels) and alcohol combine to form esters: the fragrance. Alcohol also mixes with air to produce aldehydes. Tannin is withdrawn from the wood to help flavor and color the maturing product. Four years is the average time that an American whiskey is aged, but Scotch and Irish whiskeys are generally held eight to 12 years before bottling.

Spirits and the Government

Historically, the distillation of spirits in America has always been under the watchful eye of the government. Registration of stills was required in some areas before the American revolution, but the system hardly developed wings before Treasurer Alexander Hamilton levied a tax on the manufacture of schnapps. Stills were taxed at about 50 cents per capacity gallon, and the spirits' purchaser paid about 10 cents a gallon tax. All hell broke loose as our young nation of rebels revolted. In 1794 a showdown between the Treasury Secretary's militia and the hard drinking insurgents of Pennsylvania took place, at which time the government showed that it had the power to enforce its mandates. The spirits tax stuck.

The grain whiskey makers had learned a lesson in their Whiskey Rebellion of the 1790s: They joined forces with politicians and put pressure on the government to cut off the importation of molasses; thus they cut down their rum competition. In time, whiskey manufacturers' power grew so that they influenced the repeal of the alcohol tax. The tax would be off-again, on-again for the next 50 years in cadence with the rumblings of war and temperance movements.

As the Civil War escalated, the need for federal funds increased, and the Internal Revenue Service opened its doors and demanded that the distillers pay up or close down. As in previous wars, the whiskey tax was a prime source of money. Moonshiners, men who make illicit whiskey, allegedly by the light of the moon, earned their name during the war years. Unfortunately for the federal tax collectors, many producing stills were in border states or in the south, where moonshiners thought it their patriotic duty to produce all the whiskey they could. They sold it to whoever knocked at their doors with the most money. Kentucky, with bourbon, its new corn whiskey, spread the good word about its product and for a while thrived; but like all southern industries, Kentucky bourbon was doomed to slow production because of lack of supplies.

After the cessation of hostilities, the federal tax on spirits rose to $2 a gallon and a passive resistance—the quiet nonpayment of the tax—took shape. Northern urban distillers acted as the advance guard, but the bulldog distillers of the southern hill country gave teeth to the moonshiner movement. Many people of the south saw the whiskey tax as a northern penalty levied on them for the war. They believed that the monies collected went to northern concerns, thus many southern politicians supported their tax-dodger cousins even though they believed in temperance. The Treasury Department's Alcohol Tax Unit enforcement officers became the enemy of the still operator, and citizens looked the other way. Gradually, the deadly game of Revenooers and Moonshiners became an institution. As the hunters and the hunted squared off in the war on tax evasion, federal seizures of illegal stills became meat for the media. Internal Revenue officers were held in low esteem by the people among whom they worked; some men were maimed by gunfire, some were tossed into tanks of mash. The mountain moonshiner, generally a peaceful man, became tied down by his own society. He held no hope of escaping out of the illegal corner where he had boxed himself.

In an effort to combat distillation and the movement of illicit spirits between states, Congress banned transportation of all liquor into dry states, that is, states that had voted to prohibit the sale of liquor. They also passed the income tax law at this time (1913–14), which they saw as a tool to be used against liquor-tax violators.

With the beginning of World War I, the liquor tax increased to above $6 a gallon, and Congress laid the groundwork for the National Prohibition Act, the Eighteenth Amendment to the Constitution. The Prohibition Act of 1920 made illegal the sale of alcohol other than for medicinal, religious and industrial purposes. Though rural southern moonshiners had no formal organization except fringe association with the KKK, whose members bought their product, the northern urban illicit distillation industry was highly organized. The moves to outlaw all drinking liquor played straight into the hands of northern crime syndicates. Urban distillers increased their output of the highly sugared distillate popskull. Illicit distillers redistilled legal industrial alcohol; they counterfeited tax-payment stamps and smuggled foreign liquor into the country. Often the foreign product was American moonshine that had been shipped abroad, relabeled and smuggled back.

The prohibition of legal purchase of potable alcohol increased illicit alcohol sales a hundredfold, and the bootleg industry modernized its distillation equipment and packaging and distribution systems. Some stills were three stories high and hooked into intricate antifume schemes, such as a pickle processing plant or a fish-house front operation. Some stills took up whole city blocks and were licensed as warehouses. They even had their own sewage disposal systems. Big-time bootleg operators prospered concurrently with their control of some political and financial arenas. Everyone wanted a piece of the action; speakeasies proliferated; druggists and storekeepers served as local outlets; trucks and automobiles transported the stuff; payoffs and hush money were accepted practices; and real criminal factions muscled into the whole illicit spirits racket.

Rural moonshiners also profited during the 1920s' Prohibition charade. They converted car radiators into condensing units, and bathtubs into fermenting vats. Some replaced their telltale wood fires with coke or gas burners, which heated water a distance from the still.

They ran their operation by steam. Generally, their product was a cut above the embalming fluids produced and sold in the cities. Buyers from coastal importing firms were glad to get rural moonshine even though much of it wasn't aged. They blended it with a few batches of foreign merchandise, rebottled and relabeled the booze, and sold it for a fancy price.

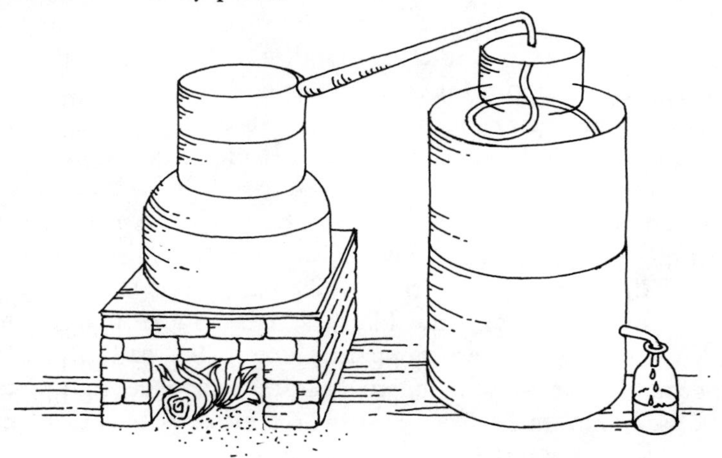

A MOUNTAIN TEAPOT: MOONSHINE STILL FROM THE 1920s

All manner of new distillation equipment and methods evolved during Prohibition's heyday. Illegal whiskey paid off many homestead notes, and in a sense helped the south to rise out of its chaotic past. As an external threat unifies a people, moonshining helped to unify America. The common threat was the Prohibition Unit of the Bureau of Internal Revenue.

Whiskey-war-weary Prohibition agents no doubt greeted the 1933 repeal of the Eighteenth Amendment with gratefulness. Theirs had been an impossible task. As a soldier does his duty, so enforcement officers did their job, no matter how unpleasant.

The transition from illicit operations to legalized control and taxation of American liquor manufacture was not an easy birthing; in fact, birth pains extended into the 1940s. With World War II threatening, the whiskey tax again increased from $2 to $6 a gallon; but there was no outcry because energies were focused and there were jobs for everyone. During the war shortages of grain, sugar and equipment made distilling difficult, young moonshiners were siphoned out of the system and bootleg pipelines withered.

By the time V-J Day's cloud penetrated mankind's collective intellect, when the horrendous potential that Einstein had unleashed sank into everyone's senses, and vibrations regarding the terrible responsibility man had taken upon himself crystallized in millions of minds, people raised doubts about the "rightness" of federal decisions. They were ready to support their own local and state authorities, and states were given a more aggressive role in policing illicit alcohol manufacture. In addition, mobility had made inroads into the solid south. Many returning soldiers aspired to training in different skills. Making moonshine appeared to be a dying art.

However, in the 1950s the federal whiskey tax rose to $10.50 a gallon, and local licensing taxes also increased. Result: Moonshining made a comeback. Illicit alcohol could be produced at less than half the cost of the federal tax alone. It did not take long for ex-G.I.s to move into the business, and their stills were a far cry from stillroom kettles or mountain teapots. They devised truck-enclosed mobile units, 18-wheelers moving innocently down the road, all the while perking out White Mule. The moonshiners devised remote-control operations. Some outfits set up networks of steps that involved one company malting, another mashing, a third operator actually distilling the stuff, maybe 200 miles away in a trailer on a remote seashore. A fourth chemical-front company rectified and bottled the illicit product before shipping it amid legitimate substances to a distributor.

Distributors were often tied with crime syndicate operations that stretched from New York to California. Even with his complex organization, the modern moonshiner of the late 1950s made fantastic money, and once again the illicit industry took off. In response, the Alcohol and Tobacco tax agents launched an attack to nip moonshiners where it hurt: in their supplies and in the courts. By 1960, tax officials computerized and put the bite on sugar and yeast suppliers and interstate transportation facilities. They also initiated an income-tax clampdown. The Treasury men conducted campaigns to catch moonshiners in the act, but their 1930s tactics of personal combat games were discarded along with the moonshiners' old radiator condensers.

By the late 1960s, the enormous profits realized by satisfying the urban poor's need for hooch defied control. In addition to the small operator, syndicate money supported the construction of huge stills, which were patterned after industrial stills that automatically performed every step, from the return of low-grade substances for rectification to the bottling of the product. Media, mobility and money promoted and/or increased alcohol consumption by all ages and classes across the country. As the youth culture, with its age-old rites of passage, swept the nation, parents shrugged and looked the other way, even as a new viper was coiling itself in and around the illicit alcohol scene. Men had been to Vietnam and had known the highs of drugs. Hashish was cheap and easy to come by, people's rights movements emboldened many subadults, and most significantly, the great society's concerns with health and the hazards of drink expanded across all class lines. Positive and negative symbols clustered, creating more and more confusion throughout the 1970s and into the 1980s. Society, led by the media, seemed intent on creating false gods of those who would break the law and get away with it.

Within this cauldron, the Alcohol, Tobacco and Firearms fricassee stewed. ATF was responsible for administering laws relating to the taxation, production and distribution of alcohol; its job has been likened to killing fleas on a cougar.

Joblessness and the wider basic education of ordinary Americans of the early 1980s has been reported to herald a rebirth of backyard stills. Make-your-own homebrewed beers had been legalized, so distillation seemed only a step away. Compared with the mammoth tax-evading crime syndicates, which were said to generate annually 75 or 80 million gallons of illicit whiskey with tax revenue loss of over a billion dollars each year, tax loss by the distillation of spirits for home consumption would seem to be nominal. To some people the stills were a curio, a naughty conversation symbol; other "family" stills were said to be fired once, like a new toy to be discarded after the novelty eroded. Truth or rumor, I cannot determine. In my circling I have not seen or heard of a backyard spirits still. Either I look suspicious, or the fact that I do not care for the hard stuff has limited my perception of the illicit critter.

Types of Stills

Although the majority of early-day small stills employed classic still-room equipment (in which sweet beer or wine is brewed, strained, heated, and the vapors directed through a water-cooled condensing tube), variations of still designs have evolved into several categories.

The simplest still for distilling spirits continues to be the pot still with a boiler heated over a direct fire and a head that compresses and directs vapors through a condensing unit.

The next simplest is the pot still with a thump keg, called a splash can or slobber box, which thumps or slobbers steam into and out of a small unit located between the boiler and condenser. This miniature container catches impurities that come over with foam, and it has a spigot on the bottom to drain watery condensates and debris. Vapors containing spirits pass through the thump keg. Some people call this contraption a doubler because if it is located very close to the boiler, the steam entering the keg heats it very hot. In effect a partial re-distillation takes place in the keg because as the vapors pass through they are reheated and concentrated before they enter the condenser.

Steam stills may be as simple as a couple of glass coffee makers, one for boiling water and the second for holding and distilling the sweet beer, plus a glass water-jacket-enclosed condensing tube. A tube transfers steam from the first to the second vessel and extends down into the sweet beer, coming within a half inch from the still's bottom which holds a few glass marbles to break liquid tensions. Steam perks into the beer and heats it so that volatile vapors boil out of the still vessel through a short, second tube exit in the cork. The exit tube connects with the water-cooled condenser.

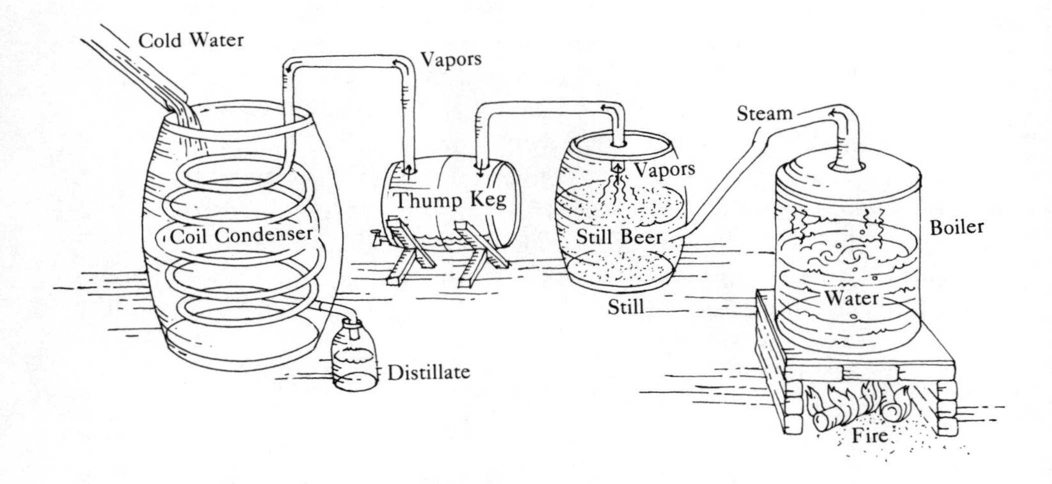

A STEAM STILL FOR ILLICIT WHISKEY

Steam is also used to distill alcohol in barrel operations. The barrels, three-fourths full of still beer, are located a distance from the boiling water. Steam is piped to the bottom of the beer barrels. As the beer is heated, alcohol and water vapors exit through a pipe to enter a thump keg. There, some vapors condense and drop, but some, predominantly alcoholic vapors, stay levitated to pass on to a coil condenser. Water-cooled condensers are generally employed to contract and squeeze liquid from the vapors. In some operations pipes carrying steam radiate like spokes of a wheel from a central steam unit that heats six or eight barrels of still beer at a time. Each barrel has its own doubler and condensing network. The distillate of such a labor-intensive setup usually runs into one collector. A multibatch steam still keeps several distillers hopping, but it eliminates the worries of tending a direct fire under each still.

There are underground illicit spirits stills called subs and groundhogs. Subs are double-wall, boxlike boilers that are built over a long ditch in which the firing mechanism is laid. Oil, butane gas or wood, all of which must extend the length of the boiler, heats the still beer, which is sometimes first fermented in the sub itself. Boiling can be tricky when done in the fermenting vat; mash residues clog condensers. These large subs are reported to produce over 150 gallons of single-run whiskey, but cleaning is difficult, and often an inferior product is the result. Whether legal or illegal, cleanliness is vitally important in distillation. An off-taste or toxicity may be caused by spoiled foreign matter. A long time ago I read of a man climbing down into a submarine-type still to scrub it. When his comrades noticed his inactivity, they investigated and found him dead. Latent carbon dioxide fumes had asphyxiated him.

Groundhog stills are built into the ground like in-ground garbage cans. When not distilling beer, only the flat, manhole cover lid is visible. The mash is fermented in most groundhog stills, and when ready, the top ferment is skimmed, the bottom sediments are drained and heat is applied by means of steam. The alcohol recovered from both the sub and groundhog stills must be redistilled in another type of still; generally it is sold raw to illicit processors.

Barrel stills were popular following World War II because the United States more or less populated many remote acreages with empty petroleum drums. I saw the remnants of a barrel still complex near Galena, Alaska. The Yukon River sternwheeler had stopped to take on wood; there was no village in sight, just the stack of precontracted, cut wood and a path meandering through the forest of shore saplings. While bargemen were loading fuel, my friend and I explored. Bouncing along the well-worn trail, which was as level as Kansas, we suddenly came across an enormous airstrip complete with acres of empty

55-gallon metal drums. We later learned that the landing field had been a refueling stop for planes enroute to Russia during the war. Near the edge of that graveyard of gasoline drums, we discovered a covey of fermenting vats, stills and stoves, all made from converted drums. The stoves were fired by wood and the stills were sealed with good old Yukon River mud. A huge tractor or earth-mover radiator condensing unit towered above the still site. We had no time to investigate further because at that moment we heard the paddle wheeler toot two blasts, the railroad signal for calling-in the brakeman. The riverboat was run by the Alaska Railroad and the signal meant that we had to sprint back to the landing. We arrived just as the plank was being lifted. Fortunately we had a tail wind and were able to leap aboard. But it would have been fun to have been stranded and to have explored that primitive country.

Some contemporary barrel still operations are said to employ heavy-duty plastic trash bags as liners for their metal fermenting drums. The trash bags are clean and disposable, eliminating the need to scrub mash barrels. Plastic is of little help in actual distillation, however. Copper, glass and stainless steel are reported to be conventional still materials. Metal such as galvanized tin or iron is unacceptable for use in distillation because of the presence of lead, which may leach into the distillate.

The Potable Spirits Law

In the study of spirits (the term usually meaning distilled liquors) it is important to understand that the manufacture of potable spirits is illegal without proper permits, inspections and in-bond tax payment as outlined in the Distilled Spirits Tax Revision Act of 1979 (Public Law 96-39). A copy of this law is available from the Department of the Treasury, Bureau of Alcohol, Tobacco and Firearms, Two Penn Center Plaza, Philadelphia, Pennsylvania 19102.

Spunkie's Method of Making Shine

Spunkie McC distilled in a copper pot still that had belonged to his father; he had real affection for the thing. At times I think he believed that the fat, bent-up cooker was his father; he talked to it, and when he showed me the place where a bear had bitten its head, he petted the patch with tenderness.

One Sunday I remember McC inviting a gang of churchgoers to a mash-making demonstration; it was really a church picnic, and in view of the circumstances I took his mash-making word as the Bible truth. However, keeping in mind that the manufacture of spirits is illegal and that time has a way of dimming mistakes, caution is advised against relating Spunkie's rough and ready skills to contemporary life.

He said that his formula was a combination of the old ways and the new; he used grain, malt, sugar, water and yeast: one part grain, one part sugar, 12 parts water, one-third of the grain weight in malted barley, plus yeast. His was a by-gosh-and-by-golly variation of his father's latter-day formula, which he said was a bushel of ground grain, that is about 60 pounds (30 kg) of ground corn, rye, oats and wheat; 60 pounds (30 kg) sugar; and about 90 gallons (360 liters) water. Spunkie's father employed the first half boiling water and the last half tepid. The senior Ira McC was reported to use about 20 pounds (10 kg) of malted barley to the bushel of ground grain, but Spunk told us that he used six three-pound (1.5 kg) cans of commercial malt to a bushel of grain. He said that on his first distillation run he recovered about 10% of the volume of the sweet beer, which was about 70 proof alcohol. On the second distillation he collected about four-fifths of the low wines, as the first-run yield is called, and that the second-run turned out to be at least 100 proof alcohol. He did not produce singlings—whiskey distilled only once—adding that it was only good for roach poison or back rubbing.

Although we did not see his still run, he explained his technique, and we all dived in to help him "wump-up his mugwump," as he called his product. I also dived into the North Pacific that day. It was Easter Sunday, a bright morning but colder than an Eskimo's ice pick. Someone had brought their cruiser filled with picnickers and had tied it to the end of the barge dock. Manning a skiff, I was helping to lighter churchgoers to land. I hopped ashore with a line to pull the dinghy up onto the beach when a giant wave bore the small craft right over top of me. I slipped, and lying flat on my backside, I was swept out to sea. I saw the skiff above me, heard people scream, but I was powerless as I skithered feet first into the icy deep. It was just not my time to be crab bait, I later remarked, because within moments, a second wave hurled me fizzle-end up, looped me over, and deposited me high, but not dry, near the spot where the dinghy lay stranded above the surf. The only discomfort I had endured was a very cold, sand-blasted tookie, but after hot coffee and a change of clothes, the mashmaking proceeded.

First Spunkie had us rinse six 20-gallon fermenting cans with boiling water; into each he had us divide the ground grain: ten pounds (5 kg) grain to each can, to which was added ten pounds (5 kg) sugar. About five gallons of boiling water was stirred into each can. The scalded grain was stirred for an hour with a stick, and when everything became a mush, we stirred in about five more gallons of hot water, got it all into a sloppy solution and covered each can. We draped extra coats, burlap and old blankets over the cans to keep them warm while we dined on moose hams someone had made, hot roasted potatoes, mulligan and luscious Alaskan strawberry cream pie.

A few of the men had been sipping moonshine, and after lunch they decided to take the dinghy and explore the exposed barge. At the time I did not know that the derelict barge was McC's aging cave. He did not stop them, rather he stood ashore and watched steadily as they tried to maneuver into the black watery cavern of the barge's upturned bow. A heavy surf ran with the tide so that at times the cockle-shell skiff seemed above our heads, then again the craft was sucked down into the swirl that seemed to exit through the bowels of the barge below. Spinning, laughing, calling, the men grabbed timbers and tried to pull themselves into the barge cavity. Suddenly, just as one man hoisted a bottle to salute his beach-side audience, the trio disappeared and immediately hollow bellows erupted from inside the bow.

Spunk's face grew long, and in two strides he was through the surf and had leaped aboard the barge. Grabbing a tub of halibut lines, he began heaving hooks the size of my hand, one after another, into the black hole. Everyone thought he had gone mad, but when he threw the other end of the line, a long rope to which the hooks and their six-foot leaders were attached, and yelled for us to run the line around a tree and start pulling, we did. The fish line with its numerous hooks was empty at first, but just as we were ready to give up, something tugged on the other end. A great shout arose as the skiff, complete with its three inebriated inhabitants, sailed forth. Again the clown raised his bottle in salute, at that moment a ground swell raised the dinghy and flattened the funnyman against the bulkhead. His goose egg was the only casualty.

When the men landed they told weird stories of being in the belly of a dinosaur, where enormous stomach worms writhed around them and the creature's backbone appeared as kegs lined up with their bungholes open.

"Bungs open?" Spunkie's gasp almost gave away his hand.

Later everyone heaved-to and completed the mashing by gradually mixing five gallons of lukewarm water into each barrel of the sweetened and scalded grain. One three-pound can of malt was added to each container. Our instructor stuck his finger in the mash to test temperature, saying that anything over 145°F (62°C) would kill the malt, that is, the enzyme action that helps to convert starches to sugars. Everyone kept stirring until Spunk came around with fresh brewer's yeast, which he added in the amount of one ounce (30 g) to each barrel. Next he arrived with bran, which he scattered over the mash barrels. The layer of bran acted as a blanket to keep temperatures stable, and it also helped to retard the loss of alcohol. As mash works, that is, as the yeast eats the sugars and give off carbon dioxide and alcohol, each tiny bubble that rises to the surface and bursts, sprays alcohol into the air. Bran, or if you were a moonshiner of the early-day plains, ground buffalo chips, is said to slow the waste of good spirits.

Yeasts are not overly happy if temperatures are below 40°F (4°C), or above fever heat 104°F (40°C), but they generate their own heat if the fermenting keg is kept covered and out of drafts.

We all thought the party was over at this point, but our host had one more surprise: seaweed. Spunkie led us across the shore rocks to a finger of water that reached inland from the unbroken North Pacific, and we discovered the head of the fjord to be filled with loose kelp. Some fronds were 20 feet long and as wide as my body. It was a spooky place; sober spruce guarded the cove of dead water, which silently pulsated amid the floating kelp. The chilly air hardly breathed, but in the distance the boom of surf against seaward cliffs throbbed, not so much against our ears as against our hearts. I was glad to get out of there, and grabbing a couple of kelp plants, I tugged them from the water and followed the gang to camp, where we heaved the slimy rubbery plants into cans of boiling water that Spunkie had instructed us to leave on the fires before our hike.

The kelp was heated through before being substituted for blankets over the tightly lidded barrels of mash. This heavy covering kept the fermenting mash's warmth in and bears out, our teacher told us with a gleam of pride in his eye. Later, the area was policed, the men went into the bushes to contribute their own bit of bear deterrent, and we all sailed back to Seward. A week later Spunkie left town to distill his prize.

Distillation is easy when everything is organized, Spunkie had told us; the drawback to the job was that once started, a distiller must keep going until the "last drap is run." That may mean a sleepless night or two.

When the sweet beer lies quiet as a gravestone with no bubbles perking through the bran and it smells like drinking beer, not sweet or vinegary, and when the "balls close smooth"—no frothy edges develop after poking a finger through the floating ecru top—Spunkie said he fired his still. Balling is the foamy bubbles atop the ferment; these were scooped off, but the sediments in the bottom of each mashing barrel were left as undisturbed as possible when the foggy beer was dipped into a bucket and transferred to the still. Some fermenters strain their still beer; Spunk said he did not, but he dippered it carefully, not stirring the bottom

Heating the still is a major skill in the art of distilling alcohol; the liquid must be kept just above the boiling point of alcohol, 173°F (78°C). The boiler should be filled three-fourths full with beer and stirred regularly while being heated. Spunkie told us that when fumes begin to rise, he placed a cookie tin of sand under the kettle, secured the bonnet, shoved the copper condensing tubes into place, and started cold water running into the worm box. Temperatures should be kept as even as possible.

Hazards may occur, like the tube clogging, the kettle boiling over, or the still blowing her head. We were advised that if equipment was kept in first-class shape and the fire tended properly, accidents were kept at a minimum. Distilling alcohol, like any other skill, takes practice and concentration in order to perfect techniques, Spunkie said. In any event, he told us that he put a large rock on the stillhead and that he stood back when he heard the first burping of vapor working its way through the tube, until after the first shot was fired. Sometimes a little foam or mash inches along in the condenser and the contraption has a habit of spitting like a camel to clear itself, he said.

The foreshot, or the first belch or dribble of liquid that comes through the condenser, should be discarded because it often contains debris from the coil. Such impurities make whiskey taste like "shine," Spunkie told us.

I do not distill spirits, so I can only relate that which I have observed years ago or have read or heard about. I do not advocate the distillation of illicit alcohol, but if you do so, remember that you are dealing with hazardous material. Danger lurks at every step, and safety procedures should never be sacrificed. I once taught a beautiful girl who had lost her hand when her father's still head blew off; my grandmother lost a distal phalanx when a head blew. Actually, she lost it during beer making, but the same pressure from fermenting substances can occur in distillation. Grandma did not see the milk-can-like head as it flew past, for it took her glasses; it is a wonder it did not remove her nose. It is not advisable to distill illicit whiskey, but if you do not intend to heed legal self-destruct warnings, be careful.

The first run of the boiler appears to take an inordinate length of time to come over, a stillroom devotee told me, but once "she begins to cackle, things take off like stampeding billy goats," the man said.

Leaks in homemade stills must be disenchanting when you realize that every squirt of vapor is a sneak thief stealing your stuff. The old manner of leak stoppage was the application of flour and water dough. It should be made ahead and be ready to plop on the steaming spot. When the patch gets soggy, you scrape it off and pop on some more, I was told. Spunk said he stuffed leaks with dry sphagnum moss, which swells with moisture. One lady told me that her parents used rags. Some distillers spend a wad on metal clamps, others use sticky tape. Stills get hot, and I would bet that tape would be the very devil to remove. The thing to remember is that you have to keep taking your little friend apart to recharge him after each run.

Once you hear the hum and have plopped on your pasties and have filled your condenser tub with cold water, one man told us it was like testing a watermelon: "You know when she's ripe to bust open." Spunkie agreed, adding that the first shot was always a thrill, like the motor of your car starting after you have fixed it.

When the first-run whiskey begins to trickle through the condenser, McC informed us that no distiller has time to admire his product; he is either dipping new beer, keeping up the fire, keeping the condenser water cold, dampening the flame with wet burlap or layering extra sand on the cookie tin. He said he laid a double thickness of muslin over the collector to strain the distillate. One lady said that her parents used chamois; Spunk's father used felt, and I read of a man filtering his whiskey through his hat.

After about an eighth to a quarter of the run had been made, Spunkie tested the distillate. If it had begun to lose its cool, slick feeling and would not flash up when a bit was flung into the fire, Spunkie told us to stop the run and not worry about every last "drap."

He lidded the collector, removed the still from the heat, cooled it enough to remove the head and saved the spent beer, by pouring it into a clean mashing container. Later, when he emptied the sweet beer from a trash can, he said he poured the spent liquid from the boiler over the mash residues. This was called slopping the mash, and with additional sugar, yeast and water, it would be used as a second fermentation.

With his first distillate tucked safely away so that it would not be spilled, Spunk said he recharged the still with new material and started a new run before the fire burned low. The successful distiller keeps going from one batch to another until all of the beer has been run. Microbes in the mash do not stop working after an eight-hour shift; if allowed to remain undistilled too long, sweet beer can deteriorate into a sour mash that even hogs won't eat.

Bad booze can make you sick as a seasick cat, and some liquor can have you pushing up daisies, Spunk warned us. Lead poisoning can create chronic illness or it can be an instant killer; the ingestion of a fraction of an ounce of lead salts can kill a person, he said. I do not think he realized that he was transgressing the cardinal rule of distillers to use only copper, stainless steel or glass when he fermented in galvanized metal. He used old Army trash cans.

I asked Spunk what was meant by reading the bead, and he replied that it was the old timers' way of measuring the alcohol content of their whiskey. It was believed that if the bubbles caused by shaking the jar of moonshine sat half in the alcohol and half out of the liquid, the distillate was fifty-fifty, alcohol and water, or 100 proof. If the bead sat low, the alcohol content was said to be less, and if the bubbles floated on top, the run contained over fifty percent alcohol.

As the sweet beer to be distilled was dipped or poured from its fermenting container, the foamy top and the mushlike bottom were reserved for a second fermentation. Spunk said he saved four things to make his second-time-around whiskey: the balls, the bottom, the spent beer and any "tailings," by that he meant leftover beer. I have heard that the liquor made from these refermented residues was called sour mash whiskey because some of the grain becomes a little sour, just as wine sours if left uncorked; souring lends a specific flavor to the alcohol produced. I have read that commercial distillers add lactic acid bacteria and other microbes to sour the mash.

To run a second-time-around whiskey, Spunk said he put all of his residues together, added cold water, half the original amount of sugar and a little yeast. In about five days he distilled it. Second, and even third-time fermentings of the same ground grain are regular distilling procedures. The later batches ferment faster and give less alcohol than first runnings. Spunkie refermented his mash only once because he said that his pigs preferred their spent mash with a little buzz in it.

Grain-based whiskeys utilize sugars and starches in their fermentation; proteins are not readily consumed by yeasts, so residues of the still contain nutrients that may be added to animal feed or applied to gardens as fertilizer. Some commercial distillers dry and press their spent mash residues into pellets and sell the high protein concentrates as DDG, Distillers Dried Grain, for animal or human consumption. Spunkie's pigs knew a good thing then they slurped it.

After he had collected all of his alcohol, washed his equipment and set his retread mash for a second fermentation, Spunkie stowed his squeezings until he had enough liquid to make the double-run batch large enough to fill a barrel. He used old Army butter kegs, which he steamed to leach salt. After they dried he charred them by laying them on their sides, placing hot coals in the kegs and turning them until their insides were burned black. Soured barrels were sweetened with a heavy sprinkling of baking soda before steaming and charring. When everything was ready he "doubled."

To redistill whiskey, that is, to concentrate or rectify it, a smaller still was employed. It perked as fast as corn popping, we were told. A man who apparently knew about spirits added that "You gotto watch her 'cause whiskey burns like sugar." For re-distilling whiskey, Spunkie said a little water was put into the bottom of the small still together with glass marbles to break up tensions. The first distillate was added to fill the still three-fourths full before a gentle fire was lit under the still. "A watchful eye creates the clearest crystal liquid made by man," the distiller said. I agreed because I remembered Spunk's mugwump sparkling as he decanted it into his prepared oak kegs. It was on an extremely low tide, the equinox moon, and he was preparing to move his whiskey to the barge.

He confided that day that he, like his father, fermented by the moon. The growing moon was nature's time to gain vigor; life reinforced itself during the first two quarters of each moon, he believed. Spunkie McC's ideas rolled from his tongue like the ground swells coming in from the Pacific. Though we were aboard the grounded barge, I became seasick. My malady always embarrasses me, but old

Spunk took my upset in stride, saying the waning moon was the time for "securing" what you have, a time for distilling. When I asked about the full moon or the dark of the moon, he cut short his enthusiasm and spit out, "They belong to the devil."

He told a story of his father, who several times had set mash to ferment on the full moon. By this time the two of us were aboard the Hell's Kettle, nosing into a strong north wind toward Seward. I can still see Spunkie McC standing straddle-legged against the lunge and slap of the bow; his angular jaw was set and his immensely strong hands gripped the wheel as he looked backward in his mind.

"My father was a solemn man, born and raised in the mountains; he was quick to sense if things was not right with a person or a drink. He never made no Cannon Ball or Swamp Muck, he was an honest moonshiner and expected other people to deal honest with him. There was a farmer who run his cattle next to my father's place, and they kept breaking the fence and eating the garden, doing all kinds of ornery things. Dad would complain, but in the end he had to run the cattle back by himself and fix the fence; the farmer didn't care. When he went up to the still one time and found it knocked over, butted, all bent out of shape, he loaded his gun for beef.

"My mother was a big woman, Polish, Dad had met her in town; she was a town woman and people weren't good to her because she didn't go to their church. Mom didn't want no trouble so she announced she was moving the family to Oregon. Her brother worked there and he wrote that lumberjacks could use some good moon. He run off, but anyway we moved, and Mom, she died, a porcupine bit her. Rabies, and my father nearly died 'cause he was all alone. Then he started making whiskey on contract; he'd deliver the stuff to a pick-up spot and he'd get cash in the mail. The men cut aged Canadian whiskey with his stuff. He didn't like the idea, but he needed the money; he had me and five others to feed.

"We was in school. Dad, he never learned to read and could only write his name, but he got me to show him how to make his letters. He'd practice writing while his still perked, he'd practice at night by the lamp, at the wood pile, in the outhouse, all the time he was writing and after awhile he learned how.

"He made pretty good on his whiskey, his contract men liked his work, he never got around, never talked to no one, they liked that, too. He just delivered his stuff to the pick-up spot and his contractor delivered his mash makings there.

"Then one year, I was about a man, 14 or 15, he started sending and receiving letters, sometimes twice a week he'd snatch the mail out of the box and go up to the still. Happy. You never saw no man so happy as when he got a letter.

"One time he grabbed his letter and said he was going up to put down some mash, and he was so excited he shook all over. 'Moon's full,' I yelled after him. 'You ain't never put down mash on the full moon,' but he run into the woods shouting that he needed to get it a-working because he had things to tend to in a week, he had to get the contract filled before then.

"My father made the stuff, carried it to the pick-up spot and he never come home. Ain't never heard from him again."

I was shocked by the abrupt ending to Spunkie's story, and ill-at-ease, I asked if he really believed that fermenting on the full moon had anything to do with his father's disappearance. There was no doubt in Spunkie McC's voice when he told me that his father set mash on the full moon three times: once when he killed his neighbor's cows, once when his wife was bitten by the rabid porcupine and the last batch.

Turning the conversation to his father's letters, I inquired if they gave any clue to his disappearance. Spunk blushed and stammered that, no, they had no significance. Taking a deep breath he told me that he found the letters; they were addressed to Gertrude, his mother. His father wrote affectionate letters to his dead wife. Over the years he told her about their children's progress, told her how he missed her; he wrote about his job making moon on contract and said he was putting by a bit to age so they could have a little nest egg when he came to get her.

Strangely, the senior Ira McC answered his own letters. He wrote Gertrude's answers. He wrote two letters in each envelope that he mailed, one from him and one to him from his wife. In the later letters, which were returned to him because he sent them to his old address in the east, he ardently planned to meet Gertrude as soon as their last child left school. So it happened, Spunkie quit after the eighth grade.

"My father drained every dreg in his cup of life, his spirit lived only to be with mother again. Her last letter made it clear that she was weeping with joy to see him.

"No, there ain't no doubt, life's ferment grows with the moon." Spunkie's very words.

Mead

Like living in Alaska, the moaning dank wilderness of northern Europe demands hearty food and drink. Soups, well populated with lentils, cabbage, roots and a clod of meat, eaten with bread, butter and a mug of honey ale, have sustained numerous generations of Europeans. Mead, made first by the conquering Norsemen and later by monastics, was a byproduct of beeswax candles. Early recipes describe mead as the fermented washings of predrained honeycomb. Soldiers were paid in this earliest of alcoholic beverages. Recalling Spunkie's warnings about popskull, I can only conjecture that the mead-bombed warriors of yesteryear probably struggled through some excruciating hangovers. Mead is now made from honey and water, yeast, nutrients (citric acid and tea) and often flavoring.

Lewis keeps bees, and from time to time he and his girls—as he calls them—make mead. We have had some superb batches. When drinking mead, one must remember that its alcohol content is close to 14%, and though it tastes somewhat like beer, it should not be chug-a-lugged.

Last year's mead was made by mixing three quarts (three liters) honey in ten gallons of body-temperature water and adding a three-pound (1.5 kg) can of commercial malt purchased from wine-making suppliers. (Malt provides the nutrients for healthy yeast life.) When the honey, warm water and malt were in solution, Lewis added a cake of yeast, lidded the container tightly and, with layers of quilts tucked around the plastic barrel, fermented the mixture for about six weeks. Mead takes from six weeks to two months to ferment, so do not tie up your favorite trash can. When the yeast action nearly stopped and only a couple of gas bubbles per minute broke the liquid's surface tension, Lewis decided to cap his mead; that is, he secured it to exclude air. Carefully siphoning the brew into sterile bottles and capping them tightly, he transported them to the cellar to mature. Do not make the mistake of inviting a gang to taste your mead when bottling it. We did one year; it tasted awful. That very same batch a year later had matured into terrific mead; it was like a teenaged punk who unexpectedly emerges into mature and splendid manhood. Mead requires patience in order to mellow into the exquisite liquor that bards raved about.

Why has mead's popularity waned? Contemporary man seems to somehow degrade everything that is not immediately useful to him.

With electricity, candle power was no longer needed, and churches adopted the use of paraffin in their beeswax candles; bees were blamed for every bug bite on man's precious epidermis. Add the massive people-push to stake a claim on every hectare of finite earth, plus the development of sugar, and mead's demise is understandable. But if you get a chance to tipple this old-fashioned drink, do so. Drink a toast to the health of the earth.

Beer

After the Crusaders returned home with samples of honey grass, that is, sugar cane, which Islamic leaders had encouraged their people to cultivate in the Middle East, honey ferments were gradually phased out, and grain-based, sugar-augmented beers took their place. Europeans liked the crisp potency of ales and beers.

Early-day beer was made essentially as purists brew it today. A cereal grain, such as barley, was soaked and sprouted—that is, it was allowed to germinate at about 68°F (20°C) for about three days. The activating of the enzymes of a grain is called malting; the malted grain was dried to about a 5% moisture content before sprouts were removed. The grain was ground so that it could circulate and come into contact with other ground grains of the brew and convert them into sugars. The next step in beer making was mashing, which consisted of mixing the malted grain with other ground grains and warm water and allowing them to interact with each other for about three hours. Meaningful relationships were developed during this time after which the mash was heated to about 140°F (60°C); the heat helped change the grain starches into more accessible forms for sugar conversion. Everything was stirred to insure contact between the malted and the plain grains. The mash was then allowed to cool to room temperature for a period of several hours in order to give the malt time to work completely before it was reheated to boiling to kill the malt enzymes. The mash was strained at this point. Later steps included boiling the liquid with hops or other bitters before cooling it again to body temperature. More sugar was added (if called for), more water, and the beer was inoculated with yeast. Lidded tightly, the brew was allowed to ferment in a warm place—about 80°F (27°C)—for about a week before being bottled and capped. Various factors gave distinctive flavors to beer: different barley blends and roasting times, using oat malt, caramel and other ingredients. The beer was matured in airtight bottles for six weeks if the brewmeister had no thirsty relatives lurking near the stillroom, then it was decanted carefully to leave sediments in the bottom of the bottle before being served. Nearly anything containing sugar or starch has been used to ferment beer: corn, rice, wheat, sorghum, sap, cassava, potatoes, soybeans or—as an oldtimer told me—a couple dozen of his old socks.

Homebrewing for family consumption is legal in the United States. Lewis is our brewmeister. He experiments with all manner of special brews, but generally we buy a three-pound (1.5 kg) can of barley malt (at wine suppliers), and add ten gallons (40 liters) warm water, five pounds (2.5 kg) sugar and a package of brewers' yeast. He lets the suds ferment for about a week in a well-covered container placed in a toasty spot, and when the gas bubbles almost stop perking (a hydrometer reading of 1.005), he bottles the brew in sterile quart-sized pop bottles and securely caps each of the 40 or so bottles. Six weeks later, if we can wait that long, we broach a bottle and drink to our good life. Clear, mellow, with a bit of bitterness, Lew's beer is a salute to the goodness of microbe friends who do the work.

The bottom-fermenting *Saccharomyces carlsbergensis* and top-fermenting *S. cerevisiae* are the most popular yeasts employed in contemporary brewing. During fermentation, yeasts that swim around then rise to the surface of the brew are called top fermenters. Bottom fermenting yeasts, generally are more at home at lower temperatures and, after remaining in suspension during active fermentation, settle to the bottom and yield fuller flavored beers. By their low-fermenting temperatures they inhibit bacteria growth.

Fermented grain brews can be divided into ales and beers. In general, ales are top-fermenting brews utilizing *Saccharomyces cerevisiae* yeasts, which work best in an environment around 68° to 77°F (20° to 25°C) or warmer, and they ferment faster than beers. Ales have a tart, mild hop flavor, are pale, effervescent and contain 4% to 5% alcohol. Cream ale is infused with distilled vanilla flavored water. Indian ale is flavored with carrot water. Weissbeer is a splendid pale ale made from wheat and malted barley, and it develops a deep, heavy foam. Stouts are dark ales, malty, sweet and fruity in flavor. They develop about 6% alcohol. Guinness Stout is fermented with a few grapes to give it the zip of wine. Porter is an old ale of England that embodies a full sweet taste, which, for my palate, needs hops. Although only 5% alcohol, this dark, heavy ale will give you a head in the morning if you do not dance it off. Sparkling ale is usually artificially charged with carbon dioxide.

Beers, technically lager beers, generally contain about 5% alcohol; they are slow bottom fermenters employing *S. carlsbergensis* yeast, and work best between 55° and 70°F (12° and 21°C). Because lager beers ferment slowly, it is important to keep the fermenting vessel (plastic trash can) well lidded. Commercial lager beers are often aged for a month or two at temperatures a little above freezing; this is said to add mellowness. Pilsners are pale, light-flavored, bottom-fermented beers that rarely contain 4% alcohol. Dortminder, often fermented using 10% rice, is a low-alcohol, mildy hop-flavored beer. Munich beer is full-bodied, sweetish, with a malt bite; it is brown and contains between 5% and 6% alcohol. Some beers such as the Belgium Lambia used to employ a special lactic acid inoculant plus a bottom-fermenting yeast. More recently, acid beers are said to be fortified with sparkling wine. Malt liquor is a pilsner, strongly flavored with hops and containing about 10% alcohol. Bock, or billy-goat beer is heavy, dark and said to be brewed with well-roasted malt, which gives it its sweet flavor. Originally, it was brewed for spring encounters with the fairer sex and was reported to be an aphrodisiac. I cannot vouch for its merit. Bock beer puts me to sleep.

Beer has been considered a tonic, and in years past medicinal herbs were often infused in or brewed in beer. Made from natural ingredients, beers have nutritive qualities, and by virtue of their hops, they sedate or relax many people.

Modern lite-light-leit beers are brewed with special yeasts that have been bred to use a lower ratio of carbohydrates to protein than most yeast strains. Brewers blend different grains to get the right ratio of mineral, carbohydrate (sugar or starch), and protein in their formulas. Thus the yeasts eat fewer carbohydrates and the beers are said to contain fewer calories than traditionally fermented brews. Most home-fermented brews contain unfermented sugars because yeasts do themselves in before they do away with all the sugars; light beers are said to have less sugar.

If you make your own beer, you get the wide-screen taste: Some beers are smooth and easy going, some are nose twisters, some eruptive and demanding, a few are lousy, but, like your kids, you take what you get and say thank you. You can always use bad beer for marinating meat. Technically there is little difference between beer, wine and soft drinks. They are made of the same ingredients: water, flavoring, sugar and yeast. Warmth is needed for fermentation, and the drink must be sealed to prevent it from turning into acid.

Wine

Wine, an almost universally accepted fermented drink, may be made from nearly any fruit or vegetable, but most grape wines are fermented from the *Vitis vinifera* kin, which is believed to have originated in ancient Egypt. White wines are usually made from light-colored crushed grapes that are pressed before fermentation. Red wines are generally fermented dark grapes; seeds and skins are removed after an initial fermentation, thus color and a stronger flavor are imparted to the red wines. Dry wines are produced from either red or white grapes that contain a high acid level. The wines are inoculated by strains of non-spore-forming yeasts, which give the wines their acridity and bouquet. Dessert wines are generally sweetened with syrups, fortified with brandy and aged differently. Traditional sparkling wines undergo a special fermentation process that allows carbon dioxide to develop after bottling; other sparkling wines are charged with carbon dioxide after fermentation and sterilization of the wine.

Briefly stated, grapes contain natural juice, sugar, flavoring and microorganisms that can cause fermentation. The yeasts growing naturally on grape skins eat the sugars of the grape and give off carbon dioxide, the bubbles, plus alcohol. After mixing and keeping everything cozy for about two weeks, the initial yeasts will be nearly expired in their own alcoholic excrement, which is generally about 12%. Homemade wine must be filtered, and specially capped to permit gases to escape but lock out air that might contaminate the beverage. Later the corked bottles are racked on their sides for settling, then subjected to several decantings into sterile bottles in order to remove sediments before they are allowed to continue their very slow, in-bottle fermentation.

Most wine makers do not rely on natural fermentation. They sterilize their juice and inoculate it with a specific yeast. A good wine yeast imparts a fruity flavor and will produce alcohol in the amount of at least 14%, sometimes a bit higher. Vintners prefer yeasts that will remain in suspension during fermentation and settle quickly. They like the heavy yeasts, ones that will not float or be easily disturbed when settled.

Although *Saccharomyces cerevisiae* is said to be the workhorse of the wine industry, dry sherry wines, producing 14% alcohol, are cultured with *S. oviformis*, or *S. chevalieri*. As sherry ages, flavor develops through chemical changes due to oxidation in which an aldehyde bouquet develops and acidity lessens. Bordeaux wines are initially fermented with a *Kloeckera* species of yeast prevalent in the Bordeaux area of France. They form a "flor," or skim, that flavors the wine. Many Sauterne wines have undergone a grape fungus inoculation on the vine causing the grape skin to split; some liquid evaporates and a concentration of grape sugars occurs. Thus natural Sauterne is a somewhat sweet wine. Acidic grapes generally yield a sour wine, but through chemistry vintners have been able to reduce acidity during fermentation.

Lewis and I make wine with wine yeasts when we have them. If fruits are ripe and we have not been to the wine making supplier, we rely on bakers' yeasts. Though these strains are bred to produce lots of gas, and thus raise a bread dough, they also work in beer and wine making.

Though there is voluminous literature on wine making, our method is casual. We cover crushed fruit with an equal amount of boiling water; allow the mixture to stand, covered, with heavy muslin or a lid for three or four days; squash and stir regularly before straining it through a cloth and adding 2½ pounds (1.25 kg) sugar to each gallon of juice. Some people double this amount; the sweetness of the fruit is the determinant. The juice should be inoculated with a little yeast—one-half teaspoon (2 ml) to the gallon—covered, lidded and kept at room temperature for about two weeks. Yeasts require citric acid and tannin, so if the fruit lacks tannin, pop in a tea bag. If the fruit is low in acid, drop a sliced lemon into the brew. After active fermentation is complete, that is, the liquid lies still, put the wine under fermentation lock (a contraption that allows no air to enter the wine). Decant into sterile bottles with tight corks after the wine has stood under fermentation lock for three or four months in a cool place. Store bottles on their sides in a cool, dark place, filter and decant a second time after several months.

In addition to the use of different raw materials, a big difference between brewed beers and wine is the time at which fermentation is stopped. Beers must be capped; that is, the air must be cut off just as the yeasts have nearly eaten all of the sugars in the brew and only tiny, pin pricks of activity (carbon dioxide bubbles) are seen on the surface of the liquid. These bubbles give sparkle to beer. Wine, which contains more sugar and thus produces more alcohol, must be put under air lock or corked just after active fermentation has stopped. That is, all of the first round of wee yeasties have eaten the sugars and have exterminated themselves.

Cleanliness is vitally important in all beer and wine making. Oddball microbes can cause off-flavor or cause home brews to fizz all over the place, and you will need a dishpan to catch your beverage as it comes down from the ceiling. "Fizzy brew can be rewarding," my former neighbor said as she pointed to root beer that had fizzed on her mother-in-law's portrait. "That bottle had more nerve than I have. For years I have wanted to spit in her eye," my neighbor giggled. She added that she was going to leave the beer on the picture, "Maybe it will grow her a mustache," she whispered.

It is advisable to chill homebrewed beverages before testing the first bottle. It is also a good idea to wrap the test bottle in a towel before opening. Exploding glass is no fun. Pour carefully down the inside of the pitcher, leaving the sediment on the bottom of the bottle. If fermented beverages are not properly sealed, an acidy liquid will develop.

Brandy and Other Distilled Beverages

Like warriors with no chinks in their armor, distilled alcoholic spirits can hold their own against acid-making invaders if the alcoholic content is 15% (30 proof) or more. Wine makers of France were long aware of the preserving qualities of distilling wine into brandy, but it was not until after the Crusaders returned with their wondrous Arabian stills that stillroom brandy became the popular means by which to drown one's sorrows.

The distilling of grape wine into brandy is as simple as distilling water, with two major differences. First, it is illegal; second, the purpose of distilling wine is to separate the alcohol and to concentrate it. Alcohol has always enjoyed water; they love to stay together. Man, a genius when it comes to drinking, discovered that there is a 39° difference between the vaporizing point of alcohol and that of water. Alcohol boils first. When you gradually heat the wine to just above 173°F (78°C) and hold the temperature between that heat and about 200°F (93°C), the alcohol boils and, hate to leave or not, rises as a vapor, leaving most of its watery associates behind. The closer the temperature comes to reaching the boiling point of water, the more water will come over into the condenser; thus the more diluted the alcohol condensate becomes. The longer the still is boiled, the proportion of alcohol remaining in the boiler diminishes. The first one-fourth holds the pay dirt, and most brandy makers shut down their stills at this time.

Directions from 1790 for making brandy state that a copper pot still with a double boiler bottom was filled three-fourths full with strained wine. The bonnet was sealed and a slow fire lit under the still. When vapors started to rise through the condensing tube and the coil began to sing, the stillman had to watch the tube exit, discard the initial slobbering, then place a collector under the tube. The directions cautioned that the fire had to be lively enough to vaporize wine yet not hot enough to cause the head to blow off or the liquid to boil over. Foam can cook in tubing, cause stoppage, and back up the pressure.

The best brandy was said to flow just after the initial spurt of foreign matter perked its way through the condensing tube and was discarded until just before one quarter of the original wine was collected. It was important to keep the fire steady during this crucial period. After the first one-fourth of the wine had run, the distiller removed his still from the heat, allowed it to cool a bit, and emptied the spent wine into a tub, where it was allowed to sour before being concentrated by evaporation into vinegar. He ran a second batch of wine through his still using the same procedures, and he kept running fresh wine until all of his wine had been distilled one time. All of the distillates that came from good runs (not burned or watery) were mixed, and the brandy was distilled a second time, or until the concentrated product was about 50% alcohol. In today's scheme of things, alcohol proof means exactly double the percent of alcohol present in a liquid: 50% alcohol means 100 proof.

Years ago, distillers wet their gunpowder with brandy to determine its "proof," that is, to "prove" that it was good brandy. If it was 50% alcohol, the powder would sputter and burn when lit; if the gunpowder sputtered and went out, the brandy was less than 50% alcohol and considered to be watered.

When newly distilled, brandy is colorless and clear. Historically, brandy was distilled from less-than-super wines in order to recover the spirits. The brandy was generally redistilled until the alcohol content was close to 100 proof. It was stored in white oak barrels for from four to fifteen years. With age, a little water from the distillate evaporates and the alcohol becomes more concentrated; it also absorbs some of the tannin from the wood and becomes amber. Fruit and wood acids, together with the concentrated higher alcohols and volatile oils, give brandy its flavor and aroma.

Many unsensational wines are distilled into brandies and flavored with syrup and synthetic essences, such as apricot, cherry, date, fig, raspberry or mint. Some brandies are given proper names. Southern Comfort is peach brandy produced in the United States. Slivovitz is plum brandy, as is Mirabelle. Kirsch is sweet cherry brandy. Himbergeist is berry brandy made in the Balkans.

According to a 19th-century recipe from my grandmother's torn book, Bramble Berry Brandy was made by taking nine pints of blackberries and infusing them, together with a half handful of rosemary leaves and a half handful of lemon balm herb, for 24 hours in nine pints of claret wine. "As you put into the alembeck, to distill them, bruise them with your hands, and make a soft fire under them." After the liquid has been distilled from the fruit and herbs it may be sweetened with sugar syrup to make a cordial, or diluted to taste with water. The recipe further advises, "Stop it close that no spirit go out."

Brandy to American moonshiners means purposefully fermenting apples or other excess fruit into jack or fruit wines, straining, then distilling and redistilling the product into a clear, potent cider. Whole apples are crushed and covered with boiling water, and sugar is added in the amount of one-half to one-fourth the volume of apple crush, depending upon the sweetness of the fruit. If sweet peaches or sour wild plums are mixed with the apples, the amount of sugar would be lowered or increased accordingly. Yeast in the amount of a spoonful to a a large crock will start fruit fermenting. "When it quits working, why, you can't let it spoil; you got to 'still' it; save the cider for a party, a funeral, or something. I know when I go, I want to be rid out in style. I want to be a-flying along in front of my neighbors who is all a-perking on cider." A Virginia mountain man spoke up after a day of apple butter making. Quietly beautiful people, Virginia mountain folk love their apple butter, sausage and cider. Next to their children, cider is their most prized product.

After a windstorm, some thoughtful soul piled a truckload of peach tree limbs in our farm lane. The peaches were big, pretty and hard as bullets; nevertheless, Lewis and I picked about four bushels before hauling the branches to the ravine. We held the peaches for about a month before they finally ripened. I canned and jammed some, then made wine by chopping and mashing the ripest peaches into plastic trash cans. After covering the mash with boiling water and adding one-fourth of the volume of the mashed fruit in sugar, plus yeast when things cooled, I covered the containers with muslin, lidded them, put blankets over them and allowed them to ferment at room temperature for two weeks.

I strained and siphoned the juice into sterile gallon jugs, put the wine under fermentation lock, and the following Thanksgiving sent the young people to the cellar for a jug. That was a fragrant and hearty wine. Had I been a distiller of spirits, I would have run a few bottles of brandy. I would have expected to recover about one-fourth of the volume of the wine, then I would have had to have found a safe place to hide the stuff for four or five years. With my luck I probably would have thought of a place like a hollow tree; a year or so later Lewis would have cut it down, ruined his chain saw and blamed me. I had better stick to my wine and my trash cans; brandy making is not my line. I have read that if you plan to age brandy in glass, a few apple or oak wood chips should be put in the jug for flavor and acid.

Sake and Other Eastern Drinks

Sake, a popular 14% alcohol drink, is neither beer nor wine. Sake is made by steaming rice to a glutenous mass, cooling the mass to body temperature and inoculating it with a mold, *Aspergillus oryzae.* After three or four days at room temperature, the pasty rice ferment, called a koji, is added to a soupy mush made of freshly boiled and cooled rice. The whole mess is stirred until everything is in solution, at which time a yeast, *S. sake*, is mixed in. A second koji ferment and more sloppy cooked rice is added two weeks later; about a month after that, when fermentation has slowed to a walk, the liquid is strained, filtered, pasteurized to retard further fermenting, and bottled. Sake is more stable than beer and less tasty than most wines; it has a slightly acid flavor and is pleasant with robust foods.

In Thailand there are more than a dozen home-fermented drinks made with rice and starters of the *Rhizopus* mold clan, plus yeasts and bacteria. Unlike the cooked fermented drinks of Japan, the Thais inoculate a watery soup containing raw, unhulled rice; they cover the brew and allow it to ferment for several weeks before straining, settling and drinking the rice beer. They often flavor this lightly alcoholic beverage with fruits or ginger root. Years ago I read that Indonesian peoples buried sealed earthenware jars of a fermented drink called oo. That concept intrigued me so I put a couple of nice berry ferments down in our spring run-off. I was going to make my own oo, but someone stole my juice. I do not recommend burying potables unless it is under your bed and you booby trap it with a cattle prod.

Rice ferments ring the Orient. Some, such as India's pachwai, a highly alcoholic beer, require three or more microbe inoculations. Rice water, coconut milk, a ti root infusion and a sugary syrup was an ancient fermented beverage used in Spice Islands festivals. Other drinks combine rice ferments with palm juice and a sweetener to create the 12% alcoholic beverages called Araks of the Far East. These drinks may or may not be distilled. Kaffir corn and millet beverages, made by cracking, boiling, cooling and subjecting the grain to several microbe inoculations, are popular fermented drinks of African countries. Like the Indonesian tempehs, these fermentation processes increase the B vitamin content of the grain. Drinking high B beers built drinkers up as it broke them down.

The great communion between spirited drinks and man's palate need not be alcoholic, soft drinks can sparkle without full blown fermentation.

Soft Drinks

Years ago natural mineral waters by reason of their sparkle were thought to hold the gift of life and springs such as those of Bath in England were meccas of the sick. Inventors tried to build containers that would maintain the mineral water's effervescence but they failed so people had to travel to the spas to drink the sparkling water. Later mineral waters were artificially made effervescent by the injection of carbon dioxide gas but it as not until the mid 1800s that yeast carbonated waters were flavored, colored, and sweetened with sugar syrup. These were bottled and sealed with a glass ball insert that was forced by the gas of the drink to lodge against a rubber ring lining a groove in the bottle neck. Since the development of the crimped edge, thin metal disc bottle cap with its cork lining, the popularity of carbonated soft drinks has become universal.

The two flavor partners: sugar and acids, such as those acids present in grapes, lemons and apples (tartaric acid, citric and malic acids respectively) have been the backbone of commercial soft drinks. As in the making of beer and wine, soft drinks are made from tepid water, sugar, flavoring (often fruit acids) and yeast. In general, four pounds (2 kg) sugar, five gallons of warm water, flavorings to taste (lemon or strawberry juice, herb decoctions or distillates, or synthetic extracts), and a generous one-quarter teaspoon (2 ml) dry yeast, are mixed in a large container and the beverage is bottled and capped immediately. The key to successful nonalcoholic carbonated beverages is to bottle the liquid quickly before the yeasties can eat much of the sugar. Leave only a thimbleful of airspace in the top of each bottle. Without air, most yeasts cannot manufacture alcohol. Lay the bottles on their sides at room temperature for four days before storing upright in a cool place. The time it takes to stir the yeast into its flavored sugary liquid and to siphon it into the bottles is sufficient to create a bit of fizz.

In nonalcoholic drinks, little change takes place after bottling. Beers and wines both mature and change flavor in the bottle, but with fizzy soft drinks, what you make is what you get. Taste your product before adding the yeast and bottling; is it too sweet? Add lemon juice. Too sour? Add sugar. Insipid? Additional extracts should be mixed into the juice before bottling.

Distilled aromatic waters go hand and glove with soft drinks. Nearly any potable distilled flavoring that lends itself to a sweet drink may be used in make-your-own pops: Anise, allspice, vanilla, cinnamon, clove, mulberries, citrus juice and peeling, gilliflowers (the ancient stillroom drink that tastes terrible), spearmint (which tastes like fizzy toothpaste), sweet birch buds and spikenard roots.

My grandfather used to make Sprossenbier out of blackspruce needles, *Picea nigra*, combined with the inner root bark of sassafras. He placed the spruce needles and slivered sassafras root bark on a rack in his copper pot still, covered the material with water, fastened the bonnet and coil and boiled like the devil. The principal constituents passed over with the steam. He boiled off about one-half of the original liquid before shutting down the still. To about five gallons (20 liters) of the bitter yet pleasantly balsamic distillate (cooled to body temperature), he added four pounds (about 2 kg) of sugar plus a scant half-teaspoon (2 ml) of brewers yeast. Granddaddy lined up the bottles on the floor below the table that held his crock of spruce beer, and I was allowed to suck the tube to start the siphon. We filled and capped the bottles, laid them on their sides at room temperature for four days and then moved them to the cave, where we stood the bottles upright. After about a week of maturation the soft drink was considered well settled; it was chilled for supper. Today we are told that sassafras contains safrol, a carcinogen.

In any homebrewed beer, soft drink or wine, there will be a sediment on the bottom of the bottle. Carefully pour the liquid into a second bottle or large pitcher, leaving the sediment in the bottom of the bottle. Thus you will not stir the sediment when you pour each glass.

Potable home distilled waters to be used in flavoring drinks should be highly concentrated either by rerunning the distillate over fresh material or by redistilling it. If flavored waters are obtained by prolonged boiling of the flavoring substance, decoctions should be filtered.

Most soft drinks may be made similarly to Sprossenbier. I had read about Heather Beer made from blooms and new heather stalks. It was said to have been a beverage of ancient Scots. I asked a Scotsman in Stromness about this, and he looked me in the eye as he replied, "It be gud fer brushin yer teeth, naw fer the gullet."

Vodka, Gin and Liqueurs

One year our potato patch overpopulated itself, and I toyed with the idea of securing a permit to perk off a jar or two of vodka. One peek at the how-tos and I settled for potato wine. Directions for the fermented base of vodka seemed easy enough, but the critical steps—removing the fusel oils and purifying the product—seemed too complicated for my simplistic leanings. "Wodka, Little Water" (that's what my grandmother called me when I wet the bed) is made by fermenting potatoes and grain with sugar, yeast and water; distilling the wine; rectifying the spirit; diluting it with distilled water and activated carbon; filtering it through charcoal; and redistilling it. The ethyl alcohol thus produced is nearly flavorless. It is cut with distilled water to produce the 70 or 90 proof alcohol called vodka. The complications of securing an AFT permit and getting rid of the fusel oil are the bug-a-boos in making your own vodka. If you suddenly find yourself potato rich, plan to spend the winter eating potato soup and buy your vodka.

Since stillroom days, man has known that the volatile oils of many aromatic plants are soluble in spiritus vin. Nontoxic plant parts have been infused in alcohol, then the product has been strained and used to flavor beverages. Today, commercial distillers mix essential oils or their synthetic counterparts with ethyl alcohol. The mixture is distilled and syrup is added to form flavorful liqueurs and cordials. But many pleasant liqueurs may be made in the home by simple infusion of herbs and spices in vodka, plus the addition of a syrup made by dissolving four parts sugar in one part water. Generally 70 to 90 proof vodka or potable ethyl alcohol is used, and a maturation period of a month or two is required. Flavored alcoholic beverages suffer identity difficulties because similar products are assigned numerous names, but counterfeit taste-alikes are fun and economical to create.

Homemade liqueurs often do not look like the commercial brands. That is because colorings are often employed by manufacturers of liqueurs. In general, the coloring is added to the alcohol or to a part of the alcohol and mixed with the beverage after infusions have been completed. If you wish to improve the color of your liqueur, add food coloring. This can be done before or after maturation.

A neighbor, who became so attached to the purebred dogs she raised she could not bear to part with them and had to move to a larger house, created wondrous cordials. Rule of thumb, she advised, was four cups 85 proof or stronger, potable alcohol (she used ethyl alcohol purchased from a liquor store), to two cups heavy sugar syrup and one-half cup whole or roughly chopped flavoring. Below are some of the varieties of cordials she made.

Anisette. Into a quart of 70 proof spirits or vodka, macerate four tablespoons (60 ml) crushed aniseed, one-half teaspoon (2.5 ml) broken cinnamon bark; one-half teaspoon (2.5 ml) crushed coriander seed; one-eighth of a vanilla bean, chopped; a spoonful of orange peel slivers without the white; and one-fourth teaspoon (1 ml) nutmeg shavings. Shake, lid tightly, and stand in a dark place for a month. When maturation is complete, dissolve two cups (500 ml) sugar in one-half cup (125 ml) boiling water, stir until clear. If the syrup does not clear simmer for three minutes. Cool. Filter or strain the alcohol infusion through muslin, mix with the cold syrup, and add one cup (250 ml) 80 proof vodka. Leave in a well-corked bottle for 24 hours, then filter or decant a second time if necessary.

Aqua Bianca is made from one quart 85 proof ethyl alcohol in which candied angelica, cinnamon stick, whole cloves and nutmeg slivers (one-half cup total) have been macerated for a month. The infusion is strained, settled, mixed with sugar syrup, (two cups) bottled and allowed to mature.

Chartreuse is a liqueur made by the monks of Chartreux. Lemon balm (*Melissa officinalis*) angelica leaves, cinnamon, mace and saffron are infused according to taste in 85 proof brandy for about ten days. Sugar syrup is added after the brandy has been filtered. Maturation for several weeks imparts the classic spicy sweetness to this elegant brandy.

Cherry Liqueur is made by mashing a quart of unseeded cherries, allowing them to stand in a warm place for four days and adding an equal amount of white wine into which sugar has been dissolved at the rate of one-half pound (250 g) per quart of wine. Bottle in a canning jar, lid tightly and allow the infusion to steep in a cool place for four weeks. Strain and filter through a washcloth or coffee filter paper, bottle, cap, and store in a cool place. If any fermentation is visible, drain off a quarter of the volume and replace it with vodka. Drink that which you take off as a sparkling sweet wine. Maturation helps this liqueur.

Creme de Coffea. In a well-stoppered vessel, infuse about 3⅓ cups (300 g) finely ground, dark-roasted, mocha Arabian coffee in a quart of 70 proof ethyl alcohol or vodka for ten days. Filter through paper. If a stronger coffee flavor is desired, repeat with fresh coffee. Dissolve two cups (500 ml) sugar in 1½ cups (375 ml) water. The sugar water may have to be heated to clear. Add the cooled syrup and a cup of 90 proof alcohol or vodka to the coffee liquor. Settle, filter, and decant into well-stoppered bottles for maturation.

When a volatile oil containing a plant alcohol is mixed with water, the mixture sometimes does not clear; thus homemade liqueurs may be a little opaque. Though it may be a little smoky looking, serve your homemade coffee liqueur in demitasse cups and try King Alphonse: Nearly fill the tiny cup with coffee liqueur and weep rich cow's cream down the inside of the demitasse so that the cream floats on top. Sipped with after-dinner conversation, this liqueur blossoms with excellence.

Curacao is a liqueur made from the bruised rind of Seville oranges macerated in one quart of 85 proof brandy. One-half cup of slivered orange skin, sans white membrane, is infused in the brandy for ten days before filtering the brandy and adding two cups of heavy sugar syrup. Maturation of Curacao generally takes several months in a cool, dark place.

Gin is a spirit distilled from a barley, rye, wheat or oats beer and flavored with the distilled oil of juniper berries (*J. communis*). Some gin producers add grapes (skins removed) to the fermenting grain and flavor with an infusion of juniper berries. An extract of angelica or coriander is sometimes added to soften the acridity.

Grand Marnier. To one quart of 80 proof potable grain spirits or vodka add three teaspoons (15 ml) food grade oil of orange, two cups of heavy sugar syrup, a half cup of honey, cork tightly and allow to stand for two weeks. Using an apple peeler, sliver several curlycues of orange peelings (sans white) into three pint bottles and decant the liqueur into them. Mature another two weeks and serve.

Juniper Liqueur made with berries of the *J. communis* (*J. virginiana*) may be toxic. It was popular several centuries ago and was said by some to be the forerunner of gin. Make a syrup of four cups (one liter) sugar dissolved in a little water. When clear, mix in one quart of juniper berries and a quart (liter) of 90 proof brandy or ethyl alcohol. Infuse for two weeks, shake occasionally, strain, filter through coffee filter paper, and bottle. Mature this bitter but satisfying after-dinner drink for a year before serving.

Mint Spirits. Infuse slightly wilted mint leaves (spearmint, peppermint, curled mint or any of the milder members of the mint family) in 80 proof vodka, one part mint to three parts alcohol, for three days in a tightly covered bottle. Shake from time to time. Strain through a cloth, filter if necessary, and mix with a prepared and cooled syrup. A heavy syrup may be made by dissolving two cups (500 ml) sugar in one-half cup (125 ml) of boiling water. Sweetening for Mint Spirits is recommended at two parts mint-flavored alcohol to one part syrup.

Orange or Lemon Liqueur. Peel six fruit, being careful to discard any white membrane. Chop rind finely. Squeeze the fruit and divide the juice together with the minced rind into three pint (500 ml) canning jars. Divide a pound (500 g) of sugar equally among the jars, and add a stick of cinnamon and a small pinch of coriander seed to each jar. Fill with 80 proof brandy or vodka. Lid tightly, shake from time to time, and macerate for two months. Strain or filter before bottling.

Most herb or fruit-flavored liqueurs hold their aromatic qualities more completely in dark bottles. Coat the inside of screw-on lids with a bit of mineral oil or paraffin so they will not stick, then tighten to retard evaporation.

Peach or Apricot Brandy. Fill an enamel pan with peeled halved or whole fruit, cover with a gallon of homemade or unfortified white wine in which two pounds (1 kg) of sugar have been dissolved. The wine may have to be heated to facilitate solution. Cool and add 1½ tablespoons (22 ml) crushed cinnamon stick and a quart of 70 proof brandy. Transfer fruit and liquid to widemouth canning jars, lid tightly and allow to steep for four days. Put a cheesecloth in a collander over a bowl, strain, let the liquid settle, filter if necessary through coffee filter paper, and decant into dark bottles with tight stoppers. Mature for a month before serving.

Transylvanian Delight. The ebullient Dr. K., a likable Blacksburg, Virginia, professor, surprised Lewis and me one evening by serving his own adaptation of a traditional family walnut liqueur that he called Transylvanian Delight. Tasting the brittle, slightly sweet, deep-flavored drink, I asked how he made the liqueur and also if there really was a Transylvania. Describing an enchanting wrinkle of earth in east central Europe, he said that people in those mountains still live in the self-sufficient ways of their ancestors. Each year families ferment fruits and carry some of their wine to a local distillery, where each family distills its own spirits. From every ten bottles distilled, one was given to the still owner and one went to the state for tax. The family grandmother took charge of the rest, doling it out at festivals or for a nip after a day of hard work. Some families infused their spirits with cherries or apricots and added sugar syrup to make flavorful brandies. Dr. K.'s family tradition was to create a unique Carpathian walnut aperitif. In Virginia, black walnuts, *Juglans nigra*, are used.

When black walnuts are the size of a thumbnail and still soft and green, pick about two dozen. Wash and cut each nut in half. Place the cut nuts in a quart jar and cover with a pint (one-half liter) of grain alcohol, 190 proof. Lid tightly and shake the jar occasionally for six weeks. Strain the alcohol, which will be dark green, into a half-gallon bottle. Make a syrup by mixing two cups (500 ml) sugar with two cups (500 ml) water, heat and stir to clear. When cool, mix the syrup into the walnut-flavored alcohol and add water to make three pints (1.5 liters). Transylvanian Delight will mature into a warm brown, bitter-sweet aperitif that will call up images of snug valley villages in "the country over the forests." It is a delight.

Throughout the world galaxies of alcoholic beverages have been created and consumed by man. Whether by desperate sin tormented or great desires frustrated, many men have trained themselves to require alcohol. On the other hand, some crystal beverages are taken for pure pleasure—the satisfactions of taste. (My old friend Spunkie, who, I heard via the mukluk satellite, ended his career as a successful hog farmer, believed in the importance of taste). Some potable ferments have assumed religious significance; whole cultures have been interwoven with symbolism accredited to beverages.

Since ancient times, pulque, made from the fermented juice of Mexican agave plants, was considered a sacred drink. Collected from plants that yield several quarts of juice daily, the liquid was carried to temples where Aztec priests fermented it with honey. Prayers were offered, oblations given and the mature and splendid drink (thought to be a gift from the sun) was sipped as a ceremonial beverage. During one period, pulque touched all lives; from birth rites to death, it was received as an act of faith and hope. Though distilled to produce tequilla, a drink that in recent years has come to be abused, pulque, like numerous sacred wines, is still called by believers, "God's love made perfect."

'STILL IN THE SKY

When the sky was lead and clouds were dust
And the Earth was paved with pain
The Sun looked down on the barren rock
And decided to 'still some rain.

So he ran some gasses into the pot
And he fastened on the head
And he perked some juice for Ole Man Earth
Who up to that time was dead.

So the swamps did come and the fish and the palm
And the planet sprang alive
And man was born and he claimed the Earth
And gave thanks to the 'still in the sky.

Man looked to the soil for his water and bread
Heedlessly wasting its leaven
'Till dead men rode through the rock-strewn land
To search for the 'still of heaven.

KIPLINGING IN THE STILLROOM

9

CHAPTER

Fuel Alcohol

"Let me give you the straight poop on fuel alcohol," a debonair petroleum engineer said to Lewis and me at a social gathering. He had overheard us talking about possibly setting up a small scale production of ethanol on our farm.

"First of all," he said, "alcohol fuels cannot compete in the marketplace. It costs more to produce a gallon of 200 proof alcohol in Iowa than it does to drill, process and ship a gallon of crude oil from Barrow and refine it." This knowledgeable man had been with Petroleum Resources in Alaska; in fact, it was through our Alaskan connections that we had met years before, before he had become a suave lobbyist in Washington. There was little doubt that he was intimately familiar with the petroleum marketplace.

"Even if you could produce fuel alcohols more cheaply than gas," he continued, "they have undesirable characteristics. [By alcohol fuels, as I was later to learn, he meant ethanol, made by distilling grain ferments, or methanol, made by the destructive distillation of wood.] Alcohol does not vaporize as readily as gasoline. It needs less air, in modern car engines it needs a ten-to-one compression ratio, or more. And it is corrosive to automotive parts which were designed for petroleum use."

"Wait. Wait," I stopped his rote, "let's go back." "I'm no auto mechanic. It takes me time to master anything more complicated than adjusting the halter on my donkey."

Our friend's handsome blue eyes lit up. He enjoyed making his point and driving it home.

"Alcohol eats plastic like your donkey eats hay," he said. "Corrosion from certain plastics and rubber is the problem. Fuel tanks, pumps and fuel lines would need constant checking and possibly have to be replaced with metal. Fuel filters, particularly in older cars, would have to be replaced over and over because alcohol is a solvent and it strips fuel system residues, which clog the filter.

"The conversion of a standard four-cycle gasoline engine from petroleum to alcohol also necessitates carburetor changes because a larger volume of alcohol than gas is required to run a car. The main fuel jets would have to be increased 50% in area; that would mean," our friend paused a fraction of a second before adding, "25% in diameter.

"Air mixture idle jets would have to be opened 25% in diameter."

"Wait. Wait, stop: It would be easier to use donkey power," I protested. He had lost me, I told him, but he had made his point. Because he said that small-scale production of methanol is not practical, I determined to research the making of the fuel alcohol, ethanol. As a home fermenter of wines and beers, I wanted to learn about fermenting and distilling fuel alcohol.

Almost every source I consulted cited different statistics relating to the energy balance of ethanol production. Options ranged from minus 260,000 BTU to a positive balance of 73,000 BTU for each bushel of corn processed. Other issues: The large amount of water needed in the making of ethanol, the fragility of storing alcohol, the complexity of regulations, the illegality of selling surplus ethanol, the limited-to-farm use and the potential earth damage of row cropping and gleaning organic residues were all questioned.

On the other hand, the supplemental use of a renewable resource, which a nation could obtain by planting and growing an ethanol fuel crop, is a sound concept. American brain cells can master car conversion hangups. Certainly Brazil's employment of ethanol as an automotive fuel is an example of "using what you got." A local farm mother used those words once in telling how she had propped her son's bed springs to patch a hole in the chicken fence caused by the son's failure to set his truck brakes.

Fuel alcohol, or ethanol (I use the terms interchangeably here), is very, very similar to drinking alcohol; in fact they might be twins if the fuel ethyl was not purposefully made stronger and poisonous. Fuel alcohol is potable spirits that have been denatured by the addition of kerosene. The Bureau of Alcohol, Tobacco and Firearms is touchy about cars drinking untainted spirits.

In my studies I have discovered that techniques have improved since Ira McC was making his mugwump. The basics are the same—mashing, fermenting and distilling—but things have advanced beyond keeping the fermenting can cozy with kelp. In making ethanol, grain is ground, mixed with water (and sometimes enzymes) and heated to break down the cell walls of the starch. Cooking the mash, cooling it and adding malt or enzymes changes the grain starches to sugar. When the whole soupy mess is lukewarm, yeasts are tossed in and the can is covered so that the fermenters cohabit in private. After three days, when the yeasts have nearly all expired in their own alcoholic juice, the beer is siphoned, strained and distilled in contraptions ranging in complexity from smoky oil drums to dazzling stainless chimneys that look like pipe organs. Fuel alcohol stills have come a long way from the copper kettle and coils of traditional stillrooms.

STEPS IN THE PRODUCTION OF ETHANOL

Conversion of starches to sugars:
- Grind grain
- Add water
- Add enzyme
- Cook
- Cool
- Add enzyme

Fermentation:
- Set mash
- Add yeast
- Ferment

Distillation:
- Distill beer
- Rectify alcohol
- Dry ethanol
- Denature and store

Other steps involve maintaining heat or steam source, testing sugar content of beer, testing pH level of beer, maintaining fermenting temperatures, testing alcohol content of still beer, maintaining distillation temperatures, testing alcohol content of the distillate and controlling ethanol storage facilities.

Farm Fuel and Gasohol Difference

Ethanol fuel for tractors should be as close to 190 proof as possible, that is, twice as strong as the hair-of-the-dog. Several redistillations or rectifications are necessary to concentrate fuel alcohol so that no more than 10% water remains. Used in converted engines, a small amount of water in the fuel does no harm. However, if your motor homebrew is to be mixed with gasoline to make gasohol, every drop of water must be removed because gas and water separate. If you have ever suffered the agonies of having water in your carburetor on a damp, icy morning and your husband's boiling point has been reached because his hat blew off and he had to chase it across the street, and he has a special meeting at work, and he cannot start the car, you understand the necessity of "drying out" your fuel. To make ethanol water-free, that is, 200 proof, chemicals or an absorbent substance must be added.

Permits

Before jumping into the experimental production of alcohol for use as motor fuel, you must learn the Rules of the Alcohol Game. First you must apply to your regional Bureau of Alcohol, Tobacco and Firearms office for the required Experimental Permit. It will be good for two years and it will authorize you to build and operate a fuel still for your own use. Your application must include an outline of general interests describing how you wish to make, denature and use your homebrewed fuel. It must also tell where you will locate your stillroom and storage facilities and how big a project your experiment will be. You will be assigned an ATF permit number and will be required to post a bond of $100 if your production of fuel alcohol is to be 2,500 gallons or less per year. A field inspection of your operation is usually forthcoming. This is not to imply that T men do not trust you; they just want to make sure that naughty things are not accidently happening around any still. Some state and local jurisdictions may also require permits.

After prolonged investigation, Lewis and I took our friend's "straight poop" advice and decided against trying to make our own fuel alcohol. Not because it would not have been an exciting adventure but because we have no cost-effective source of raw materials with which to experiment. Also, our good old farm tractor probably could not stand the purge; like an old boozer suddenly thrust into the ways of virtue, it would probably succumb to a case of the clean pipes and leak all over the place.

Though we can't participate ourselves, we believe that real respect should be paid to those who do make and use their own fuel alcohol. They are helping our country free itself from its binding dependency on foreign oil. Our country was founded by men like them, men who acted to achieve independence. Each farmer who produces his own fuel is forging a tiny link in the chain of American independence from foreign oil.

Fuel Stocks

In investigating ethanol production I found that sugar cane or beets, grain such as wheat or corn, and cellulose from wood and paper scraps or crop residues are the three most common sources of fuel stocks. Of all the raw materials that are used to produce fuel stocks, sugar is the easiest to process. After extraction of sugary juices from cane or beets, simple fermentation and distillation are employed; sugar crops need neither malting nor enzyme treatment. Grains, vegetables or fruits are the second simplest fuel stocks; processing and harvesting equipment is minimal and techniques are standardized. Cellulose can be complex to handle; the fiber of plants must be broken down through acidic or enzymatic treatment that decomposes the plant substances into fermentable sugar.

ADVANTAGES AND DISADVANTAGES OF COMMON FUEL STOCKS

Type of Feedstock	**Processing Prior to Fermentation**	**Principal Advantages**	**Principal Disadvantages**
Sugar Crop: Cane, beets, Jerusalem artichokes.	Milling to extract sugar.	1. Minimal preparation. 2. High yield. 3. Residue valuable as feed.	1. Loss of sugar in storage. 2. Cultivation not widespread.
Starch crops: Grains and tubers.	Milling, liquefaction and conversion to sugars.	1. Storage is simple. 2. Cultivation widespread. 3. Residue valuable.	1. Equipment, labor and energy costs. 2. Residue from contaminated grain unusable.

Yield

People always want to know how much fuel grade alcohol they can expect from corn or potatoes or other stocks. The literature states about two and a half gallons per bushel of cereal grains and a little less than a gallon from a bushel of potatoes. Farmers with less efficient stills recover less alcohol than the books state; I remember Spunkie McC distilled just a little over a gallon of alcohol from a bushel of grain. The amount of alcohol recovered depends upon the fermentable sugar or potential sugar of the fuel stock. Another consideration in ultimate yield is the protein content of the material left over after the removal of alcohol. Spent mash in the fuel alcohol business is called stillage, and it can be dried to form distillers' dried grains (DDG), a valuable feed for animals. A high quality protein food for humans can also be produced from stillage, according to research done at the University of Wisconsin.

Listed below are various fuel stocks and their potential alcohol yield as based on average fermentable sugars in the material. Alcohol yield depends not only on sugars but also on the efficiency of techniques employed: temperatures, pH, the yeast strain, contamination by bacteria, distillation equipment, the condition of the substance used, and so on.

POTENTIAL ALCOHOL YIELDS OF VARIOUS FUEL STOCKS

Material	Alcohol Yield	Protein of Stillage
Apples	14.4 gal/ton	
Barley	2.0 gal/bu or 79.2 gal/ton	
Beets, sugar	20.3 gal/ton	20%
Buckwheat	83.4 gal/ton	
Carrots	9.8 gal/ton	
Cheese Whey	1 gal/44 gal whey	
Corn	2.5 gal/bu or 85 gal/ton	30%
Dates, dry	79.0 gal/ton	
Figs, dry	59.9 gal/ton	
Grapes	15.1 gal/ton	
Jerusalem artichokes	20.0 gal/ton	
Molasses	69.0 gal/ton	20%
Oats	1.0 gal/bu or 63.6 gal/ton	
Peaches	11.5 gal/ton	
Pears	11.5 gal/ton	
Pineapples	15.6 gal/ton	
Plums	10.9 gal/ton	
Potatoes	1.4 gal/cwt or 22.9 gal/ton	10%
Potatoes, sweet	34.2 gal/ton	
Prunes	72.0 gal/ton	
Raisins	81.4 gal/ton	
Rice	79.5 gal/ton	
Rye	78.8 gal/ton	
Sorghum, grain	2.6 gal/bu or 79.5 gal/ton	30%
Sugar cane	15.2 gal/ton	
Wheat	2.6 gal/bu or 85 gal/ton	36%

Most farmers whom I have consulted about fuel alcohol ferment grains shrugged when I asked them how much fuel could be produced per acre. It depends upon the yield, which depends upon the soil, fertilizers, tillage, the seed, weather, equipment efficiency, storage, handling and of course a person's skill. Most don't bother with figures. One man who does told me that he gets about two gallons of alcohol from a bushel of corn with his distillation techniques. Multiplying that by a rounded 100 bushels of corn per acre, he figures he gets about 200 gallons of alcohol per acre. He ran his tractor, picker and other implements on alcohol so he hadn't needed to "dry out" his fuel, which is an expensive operation. He seemed pleased with costs of producing ethanol, but other farmers just turned down the corners of their mouths and shook their heads. There are so many variables that cost becomes a very personal evaluation. Most farmers said fuel alcohol was not their main crop, that they only converted crop surplus or damaged crops to alcohol; a few reported buying damaged material for their stills.

Converting Grain to Sugar

In preparing feedstocks for distillation, grain starches must be broken down because yeasts can only eat sugars. The grain must be ground to the size of sand to allow water to reach all parts. After preparing a slurry (mixing the meal directly with water and stirring to prevent lumps), the pH has to be tested. The pH scale is based on numbers from 1 to 14 with 7 being neutral. Figures above 7 indicate an alkaline substance; those below 7 are acid. If the pH of the slurry is too low, or acid, add a solution of baking soda or ground limestone; if the pH is too high, or alkaline, add vinegar. Just the tiniest bit of either additive mixed into the slurry will raise or lower the pH. Testing equipment, which may be as simple as slips of paper, is available at laboratory supply houses.

When the slurry is adjusted to the recommended pH, the initial enzymes are added. Enzymes are similar to the catalysts found in sprouted or malted barley in that they assist in changing the grain starches into sugars. They are commercially available and are generally introduced into a little warm water before being added to the slurry. Then the slurry is stirred for a short time before being heated. Gelatinization of the grain occurs at around 150°F (65°C), and the temperature is held at the enzyme manufacturer's recommendation for two or three hours to let the enzymes work. Different commercial enzymes require different temperatures, pH and times for optimum activity.

Some distillers introduce a second glucose-producing enzyme after the foregoing starch conversion step is completed. This second round of enzymes catalyzes and enhances sugars at specific temperatures and pH depending upon the variety selected. Most sugar-converting amylase enzymes are said to work best in temperatures between 122° and 140°F (50° and 60°C), and the mash is stirred to assure maximum contact between the enzymes and the slurry or mash. When the grains have been changed to sugar, the mash is strained.

Converting Sugar to Alcohol

Now enter the yeasts. As in the fermentation of foods and beverages, strains of the genus *Saccharomyces* are widely used for fermenting fuel alcohol. The favorite yeast seems to be *S. cerevisiae* because those microbes jump in and work like crazy. Their frenzy discourages contaminating bacteria; they also eat most sugars, thus producing the most alcohol in the least time. In general, yeasts added at about three ounces (100 g) per bushel of corn and held in comfortable environments (between 86° and 95°F—30°C and 35°C—and pH of 3.0 to 5.5) will prosper. If the sugar content of the beer is too high, osmotic pressure can inhibit the growth and even destroy yeast activity. Yeasts are cellular organisms, and sugar concentrations greater than 22% by weight can create excessive pressures that literally rupture the cell walls of yeasts. They can also be killed by high alcohol concentrations; alcohol levels greater than 12% to 14% by weight kills most yeasts before they have consumed all of the sugar. That is a waste, thus it behooves an ethanol distiller to test sugar content with a hydrometer; if it is too high, he must add water. He must also test the yeast-inoculated liquid periodically with a proof hydrometer to keep tabs on alcohol conversion. Fermentation goes through four steps: Initially there is little alcohol production while the yeast becomes acclimated; second, the yeast cells propagate rapidly and alcohol content shoots up; third, a leveling off phase slows alcohol conversion; and fourth, the alcohol concentration is high and the sugar content of the beer is low. The mixture is about 12% alcohol and 88% water and will not burn. The alcohol must be concentrated by distillation.

Fermenting Techniques

There are as many ways of making mash as there are mashers. Some fuel alcohol producers do not like the swift demands of commercial enzymes, pHs and temperature controls. They prefer to eyeball, touch and taste-test like Spunk and his father. One contented farmer told me he uses three parts water to one part ground corn by weight. Another man uses four parts water to one part ground corn by weight. My contented farmer heats the water to around 140°F (60°C) and slowly adds the ground corn and part of the malt, stirring continuously. After about 30 minutes of "premalting," he boils his mash for about an hour in a cutoff oil drum. He stirs it from time to time. Mash likes lots of attention; it either spits at you, sulks or boils over if you do not mind it.

This farmer, as might be expected, uses dry powdered malt instead of commercial enzymes. Although the amounts of malt others use vary from one-tenth to one-fourth of the weight of the solids used in preparing the mash, my farmer friend uses about one-fifth of the weight of his ground corn. He adds half of it before the boiling step because he believes it helps the sugar conversion; he adds the other half after the mash has cooked and cooled to about 140°F (60°C). Some men add all the malt at the end of the cooking. By either method, the mash must be cooled to around 140°F (60°C) and the dry malt first mixed in a little warm water to get it into solution before adding. The mash is covered tightly and allowed to "make sugar" for about four hours.

When the time is up, the contented farmer takes a spoonful of mash and adds a bit of regular iodine to it. If the sample darkens, the malt needs more time to do the work so he stirs it a little and lets it stand, covered, for another hour. Starches darken when mixed with iodine, but sugars do not react to iodine, he told me.

It surprised me to hear that he uses a hydrometer in testing the specific gravity of his sweet beer. The hydrometer measures the amount of dissolved sugar in a liquid. Sugar makes liquid heavier, thus the specific gravity increases as the amount of dissolved sugar increases. A hydrometer resembles a glass thermometer with a weight on the lower end and a numbered scale ranging from 1.130 to 1.000 at the top. The hydrometer is floated in a jar containing a sample of the liquid from the mash. The point at which the liquid cuts across the stem of the hydrometer is the reading representing the sugar content of the liquid.

Water has the specific gravity of 1.000; thus if you were to test water, the hydrometer would sink in the liquid and only the top-most reading, 1.000, would be seen. As the sugar content in the liquid increases, it makes the liquid heavier and the hydrometer rides higher.

The higher the hydrometer pokes out of the liquid, the greater the amount of sugar present in the mash.

For best yeast participation in the making of ethanol, about 15% or 20% dissolved sugar in the mash is recommended. On a hydrometer, that equates to a specific gravity reading of about 1.060 or 1.080. If the reading is above 1.080, the farmer told me he adds water to the mash. If it is lower than 1.060, he adds sugar water to the mash; he doesn't want his yeasts to go hungry. Snockered microbes do not produce alcohol either, he added with a wink.

After the mash reaches its proper level of soluble sugar, my friend strains it through a bed sheet or a tight nylon mesh. Then he charges the milky beer with yeast. The tricky part of brewing is to keep the beer free of contamination; infection by bad microbes cuts down on alcohol yield, so working beer should be kept warm, around 80°F (26°C) and covered. Most brewers, as I have reported, use the brewer's yeast strain *S. cerevisiae* because that species can effectively utilize 20% sugar solution and produce at least 12% alcohol (baker's yeast rarely generates more than 10% alcohol).

Brewer's yeast comes in large packets containing about 50 pounds (25 kg) and should be kept refrigerated. You can order both yeast and malt enzymes from Anheuser-Busch, Inc., 721 Pestalozzi Street, St. Louis, MO 63118, or from local distilleries; also your state alcohol fuel office can put you in touch with other sources of supplies.

Brewers usually mix the yeast in a little warm sugar water before adding it to the strained liquid and fermenting it for three or four days. My informant tested his still beer on the third day, and if the hydrometer reading was 1.000 he fired his still. He reminded me that he tested his mash two times: the first after the mash liquid had been malted to determine the percent of sugar in his beer, and the second after fermentation had slowed and he needed to know if the yeasts had eaten all of the sugar. Some brewers employ a proof alcohol hydrometer, an instrument that determines alcohol content in still beer; they also use it to measure the proof of distilled ethanol.

Fermentation Locks

Several men whose operations we visited showed us their 55-gallon metal fermenting drums, which were lined with heavy plastic inserts and filled three-quarters full of still beer to allow the brew to form a head. We were surprised to find fermentation locks attached to the drum lids. They looked like clear plastic gooseneck sink drains poking up from the tops of the vats. Gas from the fermenting beer could exit through them, but no air could get in because water lay in the bottom of each gooseneck. We have had no problems with microorganisms contaminating our homebrew, but fuel alcohol men don't take any chances. Lewis pointed out that it was a good idea to use fermentation locks considering the amount of money they have tied up in their operations. If we lose ten gallons of beer to bad microbes that's nothing compared with the loss of several hundred gallons of fuel ethanol ferments. Vinegar is a one-way trip, as one man commented.

We saw fermentation locks that were sounding off like a bunch of frogs with hiccups. Others burped lazily. When the water in the locks stops bubbling, fermenters become distillers and go into orbit. They dip, pour or pump the beer into a still and do not slow their pace until the last barrel has been run.

Distilling Fuel Alcohol

The distilling of fuel alcohol is a little different from the distilling of spirits, it is more of a science.

In farm ethanol production two types of stills predominate. One is the unit processing type of operation, which is simply distilling one batch of still beer at a time and repeating until "everything, including the old cow's teat, has been squeezed dry." The second is the continuous processing of beer in a still that employs a gadget to regulate a jet of beer into the still. Alcohol is continuously separated from water and automatically redistilled until a finished product streams out the far end. Most do-it-yourselfers distill ethanol in batches. Some batch-distillers work only with converted 55-gallon drums; some use 1,000-gallon tanks for fermenting their grain and then distill the beer in 500-gallon stainless steel systems.

Fuel alcohol studies have indicated that a practiced distiller can make about a gallon of farm fuel alcohol, 160 proof, from about 15 gallons of fermented mash. By filling a 55-gallon still about three-fourths full of beer (about 40 gallons), the distiller can recover up to 2⅔ gallons of ethanol per double run. If a farmer uses 500 gallons of fuel alcohol a year, nearly 200 runs will be required. Obviously a farmer cannot spend that much time brewing and distilling alcohol.

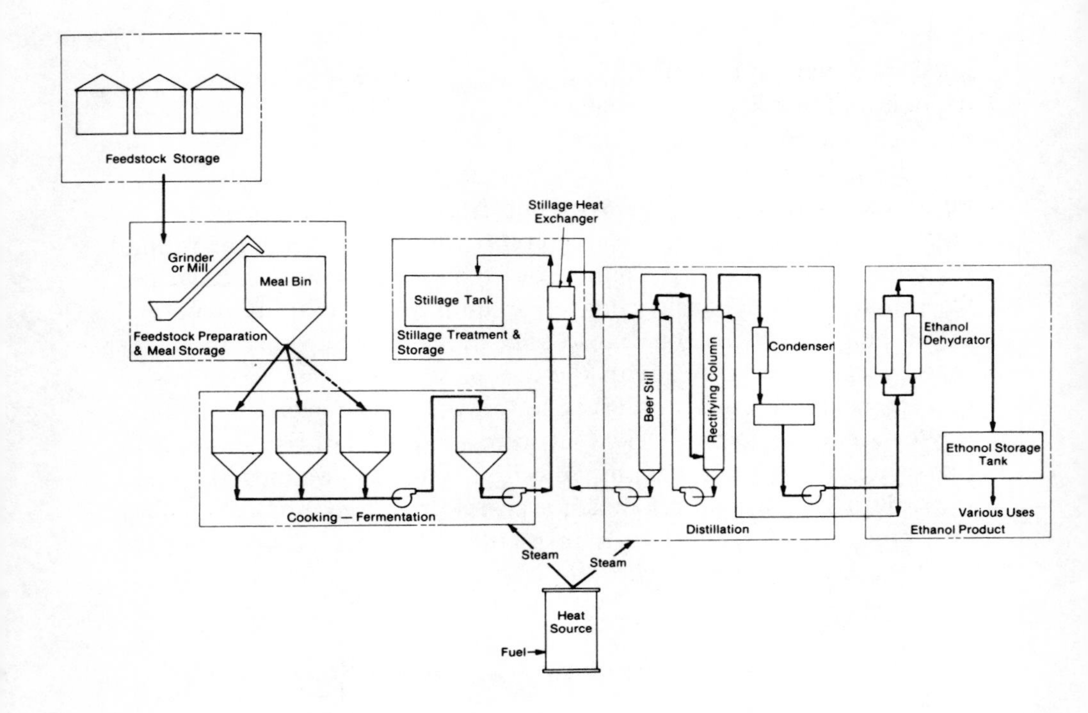

AN ETHANOL CONTINUOUS PROCESSING STILL

Deciding on the size of the fuel still is the most difficult decision, one man told me. Too big a capacity is a waste because, according to the law, a farm fuel producer may not sell or give away surplus ethanol without a complicated ATF permit.

The simplest still we saw consisted of a boiler with a thump keg and a condenser coil; it looked to all the world like pictures of moonshine operations I had seen. However, the two young enthusiasts who ran it had their permits posted and were very much "into" fuel, as they say. When the distillate coming over into the collector showed up weak, the still was closed down, cooled and recharged with beer. Over and over the young men fired their still and ran the distillate until they achieved a 160-proof, fuel-grade ethanol.

They worked like beavers, chopping wood, hauling corncobs given to them by their father, dipping fermented beer into their 55-gallon drum still, tending the fire, clamping leaks, checking condensing temperatures, testing the distillate, and pouring spent residues into other drums for concentration. And they were elated when they figured they had recovered 33 gallons of 160-proof ethanol from 20 bushels of donated corn. I have seen more efficient stills but none that provided the learning of that simple contraption. The miracle was that they were able to produce fuel grade alcohol.

They told me how they were going to improve their next still. They planned to make a distilling column out of an old water heater. The vapors would be directed from the drum boiler into the column about midway up the still, then the vapors would continue up and out the top into a condenser. Water would drop to the bottom of the column where it could be emptied by opening a tap and reprocessed if it was rich enough. Each redistillation wastes time and energy, they said, even if the heat energy source is corn cobs.

The young men's idea for using an old water heater is a kind of adaptation of the reflux, or column still, that is employed by many small-scale farm ethanol producers. It is essentially a simple still with an attached six foot or taller column, which carries the boiling beer vapors upward so that they will hit large cool surfaces and condense. Water falls, vapors rise.

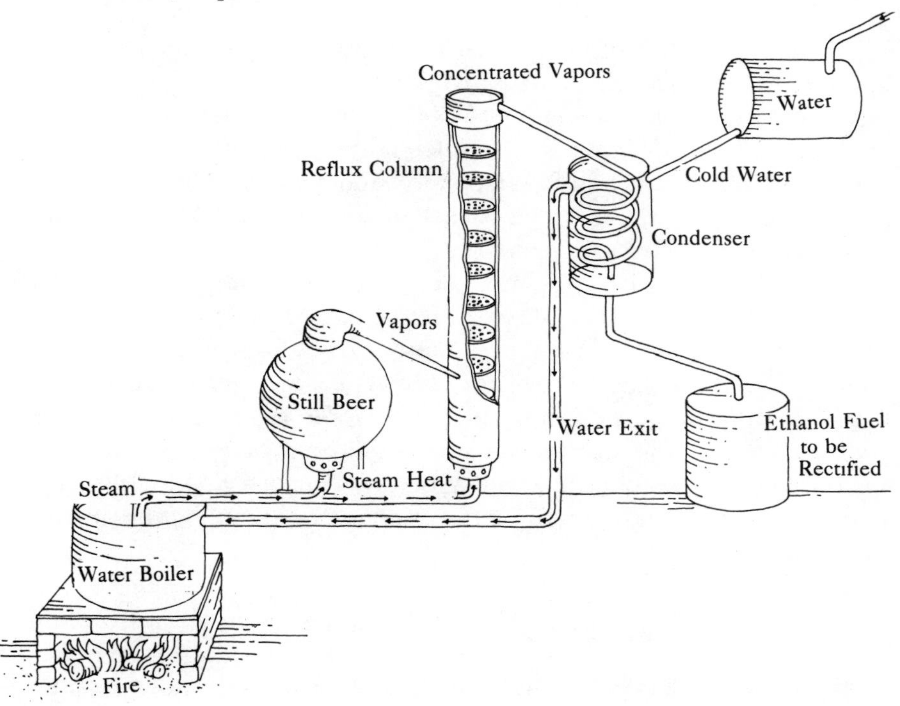

STEAM STILL WITH REFLUX COLUMN

The column still consists of a bottom, or adjacent, steam-heated beer boiler with a tall chimney attached to it, plus a condenser. The vapors from the steam-heated beer are directed into the column that is filled with loose, nonabsorbent material such as ceramic or glass marbles. As the alcohol and water vapors rise, they hit the cooler surfaces and some of the vapors condense into droplets. The water in these droplets drips to the column's bottom. If the column has been properly insulated, the inside temperature stays above the vaporizing point of alcohol and any condensed alcohol droplets will revaporize and continue upward. Again and again as the vapor hits the marbles it forms droplets, the water runs down and the alcohol revaporizes and goes up. The heat inside of the column is not allowed to rise to the point of vaporizing water; it is kept just hot enough to revaporize the alcohol after it hits the cooler packing material and tries to run down with water. Finally, after numerous condensations and revaporizations, the alcohol vapors, which become more and more concentrated, make their way to the top of the column and are rushed to the condenser.

In boiling a batch of beer, whether it is in a simple pot still or in a column still, that which boils first is the alcohol plus any volatile oils in the beer. The part that makes it to the top of the column early in the distillation process will produce a higher alcohol proof, sometimes as high as 180 proof. When the vapors are mostly steam, the still is closed down, the boiler is recharged with fresh beer, and the process of distillation is repeated.

Proper insulation and maintenance of the still help control temperatures throughout the columns. Temperature dictates the success of reflux distillation. The thermometer at the top of the chimney should be held at about 176°F (80°C). If it is hotter, steam will go over into the condenser; if it is much cooler, the vapors may condense prematurely and the alcohol will run down the inside of the column. Distilling alcohol is not unlike living with teenagers: You have to stay flexible, ready to jump, change directions, stop dead, or charge at a moment's notice.

Large Ethanol Still

Most large farm alcohol manufacturing units consist of grain storage and milling bins, cookers with automatic enzyme dispersal systems, strainers, fermenters, a heat exchanger, a steam-generating boiler, at least two distillation columns, a condenser, rectifier, drier, denaturing unit and storage tanks. Grain is milled and mixed with water; enzymes are added; the mash is cooked and cooled; more enzymes are introduced; then the mash is strained, the liquid charged with yeast, and fermented; and still beer is pumped into a holding tank in preparation for distillation. The first column into which the still beer is pumped is called the stripper; it is heated by steam and filled with baffles, plates or glass balls. A spent beer outlet is on the bottom. The second steam-heated column is called a rectifying column, and alcoholic vapors from the stripper are allowed to enter at the bottom and make their way upward; all the while they are being concentrated as the heavier, water-laden material drops. Temperatures are maintained just above the vaporizing point of alcohol. The vapors finally exit through the top and are piped to a small condenser, then to a small rectifying still and to the drier if the ethanol is to be added to gasoline to form gasohol. If the product is to be used as farm fuel, the ethanol goes from the condenser to a denaturing unit in the storage tank. I have read of multiple column stills where temperatures are graduated and the product is 190-proof ethanol.

The production of ethanol in a continuously operated, anhydrous (water-free), 25-gallons-per-hour plant involves grinding the fuel grain once a week and filling the meal bin. Meal from the bin is mixed to make a mash in a 4,500-gallon cooker-fermenter with a hydraulic agitator. Generally in a large plant there are three cooker-fermenters: one for starting the slurry and adding enzymes for sugar conversion, the second for fermenting, and the third for finishing the brew and pumping out still beer. The use of each cooker-fermenter is rotated. Modern fermenters can be emptied, cleaned and restarted without having to wait until the still drains. Beer is fed from the finishing tank through a heat exchanger to the top of a column still that is 18 feet high (six meters) and about a foot in diameter. This first still is provided with sieve trays or a stripping section that removes the downflowing alcohol from the beer by means of sieve plates that separate liquid and vapor. Vapors rise and are directed from the small head that sits atop the column to the bottom of an alcohol concentrating still that is 24 feet high (eight meters) and about a foot in diameter. This second still rectifies the fuel vapors to 95% ethyl alcohol. The product is condensed and pumped to a dehydrator.

Commercial ethanol distillers dry out the last traces of water in their alcohol by solvent extraction. They add chemicals, such as benzene, that remove water. I have read of people removing water from alcohol by mixing it with unleaded gasoline, or dehydrating it with corn or lime, but I have not seen any of these drying processes at work. Drying is necessary because 190-proof alcohol is about the best that any distillation process can attain. At about 96% alcohol, a constant boiling mixture of alcohol and water forms. Chemists call this mixture an azeotrope, meaning that it cannot be separated.

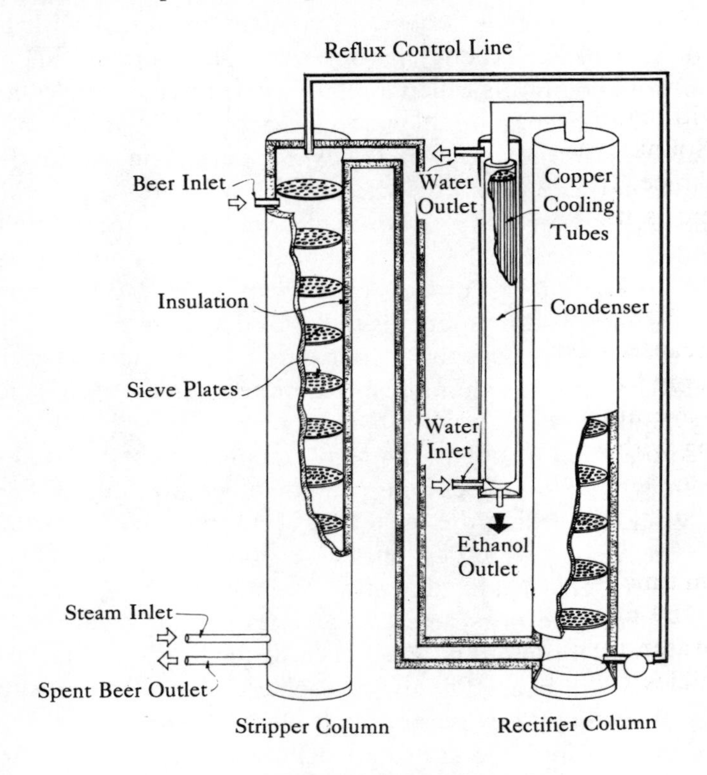

SIEVE TRAY COLUMNS

Maintenance of a large ethanol distillation plant includes daily removal of ash from burners, blowing of steam lines, checking hydraulic systems for leaks; and every week backwashing the water softener, checking the roller mill for roller damage, cleaning out back pressure pipes, sterilizing and washing down the beer cookers, cleaning the beer columns and checking pump seals. Each month all fans and belts must be lubricated and checked, the air filter of the yeast tubs must be changed, the fermenters must be sterilized, the condenser descaled and the motors checked. That sounds like a lot of work; however, modern ethanol production plants have automatic feedback controls and miniature computers that regulate flow, temperatures, pH, enzyme and yeast additions as well as timers for each step.

Processing varies with different equipment and ingredients. Some enzymes are added while cooling the precooked grain starches, some are added prior to cooking, and some are added before and after cooking. Cooking time and temperatures also vary. Some producers suggest the injection of steam into the mash; others recommend direct heat with constant agitation.

As Spunkie told us years ago, "You pays your money and takes your choice. Ferments are carefree creatures."

There is more uniformity in the practices of storing ethanol. Although not as explosive as gasoline, alcohol is inflammable; it is a dangerous liquid and there are usually ordinances controlling its storage. Storage tanks must be lidded because alcohol absorbs moisture and because alcohol evaporates. The two young men with the pot still learned about evaporation when one batch shrank by as much as a cup overnight. I think they exaggerated, but I have learned by working with fragrances that anything you can smell is losing something to evaporation. Storage tanks must also be vented because alcohol builds up pressure as it heats and cools—enough to rupture a tank.

From time to time alcohol is tested for its proof in storage. Anything below 160 proof is not considered fuel ethanol; authorities say that some water, up to 10%, improves the energy output of farm machinery or vehicles converted to run on fuel alcohol; but if more than 1% water is present in gasohol, the two fuels will separate. For use in gasohol, as said earlier, fuel alcohol must be water-free.

Federal and state laws encourage the production and use of ethanol both as a farm fuel and as a gasoline/alcohol blend for motor vehicles. In addition to U.S. Small Business Administration guarantees to underwrite the building of certain fuel alcohol plants, significant federal tax incentives are offered to the producer of alcohol fuel, and many states exempt gasohol from their motor fuel tax. The federal government also provides a solar energy information data bank, the SEIDB at 1617 Cole Boulevard, Golden, CO, 80401. SEIDB offers free guidance and literature to anyone interested in the small scale production of ethanol. Seminars and hands-on workshops are periodically offered in different geographic areas; the National Alcohol Fuel Producers Association, P.O. Box 686, Colby, KS, 67701, can be a helpful source of information. One can also get answers to questions from the National Technical Information Center, 2585 Port Royal Road, Springfield, Va., 22161.

Solar Distillation of Fuel Alcohol

The best way to get the feel for fuel alcohol production is to take a guided tour of an operating still. I often learn as much from other participants as from the tour guide. During a farm industry tour highlighting a sieve tray column still, I met an engineer who was gung ho for solar distillation of fuel alcohol. He believed that in the future huge solar collectors would be built to extract the alcohol from beer. He drew sketches. To my astonishment they resembled the distillation devices in which ancient Egyptians had concentrated their beers a thousand years before Christ.

A flat vessel containing beer was set in a sealed box in the sun. The top of the box consisted of a clear glass panel slanted upward to catch the maximum rays of the sun. As the temperature in the box began to rise, the alcohol in the beer vaporized. When the vapors struck the cool glass, they condensed on the glass, forming droplets that ran down the 45° angle of the glass into a collector. A solar still! A 3,000-year-old idea as new as tomorrow.

AN EGYPTIAN BOX SOLAR STILL

The circle is complete. The princely art of distillation has circled the sciences until again it is knocking at the door of simple sunlight, the greatest distiller in all the universe.

Throughout the evolution of stills, one enduring sentiment emerges: man's faith in the earth. From his desert beginnings man has consistently returned to the soil to look for his pure water, his helpful microbes, his flavorings, fragrances, and healing potions, his spiritus drinks and, more recently, for his distilled fuels. Over and over again, man has demonstrated his faith that our planet will forgive human transgressions.

The beginning of faith is said to be self-sacrifice and humility. To substantiate faith that the earth will heal itself, man must overcome any notion that he controls life's processes. He must sacrifice for the love of the earth.

As laughter is the bubbly distillate of joy, and rain is the distillate of that glorious still in the sky, so faith is the distillate of tomorrow.

The still is more than a tool employed to expose the splendors that lie hidden in this planet. The process of distilling affirms the entire underlying order that sustains this noble earth.

BIBLIOGRAPHY

American Medical Association, The Council on Pharmacy and Chemistry. *New and Nonofficial Remedies*. Philadelphia: J.B. Lippincott, 1954.

Askinson, George William. *Perfumes and Their Preparation*. New York: N.W. Henley and Co., 1892.

Bradley, Jack. *Making Fuel in Your Back Yard*. Wenatchee, WA: Biomass Resources, 1979.

Bramson, Ann. *Soap, Making It, Enjoying It*. New York: Workman Publishing Company, 1975.

Brodie, Edith P. *Materia Medica For Nurses*. St. Louis: The C.V. Mosby Co., 1930.

Carley, Larry W. *How To Make Your Own Alcohol Fuels*. Blue Ridge Summit, PA: TAB Books Inc., 1980.

Carr, Jess. *The Second Oldest Profession*. Radford, VA: Commonwealth Press, 1978.

Coon, Nelson. *Using Plants For Healing*. Emmaus, PA: Rodale Press, 1979.

Courtney, W.S. Revised and enlarged by George E. Waring, Jr. *The Farmers' and Mechanics' Manual*. New York: E.B. Treat, 1874.

David, Elizabeth. *Spices, Salt and Aromatics in the English Kitchen*. Baltimore, MD: Penguin Books, Inc. 1970.

Donnan, Marcia. *Cosmetics From the Kitchen*. New York: Holt and Co., 1972.

Douglas. *The New Dispensatory*. London: Printed for J. Nourse, Bookseller in Ordinary to his Majesty, 1770.

Eberle, John. *Notes of Lectures on the Theory and Practice of Medicine*. Philadelphia: Grigg and Elliot, 1840.

Fairley, James L., and Kilgour, Gordon L. *Essentials of Biological Chemistry*. New York: Reinhold Book Corp., 1966.

Fernald, Merrit L., Kinsey, A.C., and Rollins, Reed C. *Edible Wild Plants of Eastern North America*, revised edition. New York: Harper and Row, 1958.

Gibbons, Euell. *Stalking the Healthful Herb*. New York: David McKay, 1970.

Gildemeister, E. and Hoffman, Fr. *The Volatile Oils*. 2nd Ed. Translation by Edward Kremers. Volumes I, II and III. New York: John Wiley and Sons, 1922.

Gray, Asa. *Manual of the Botany*. New York: American Book Company, 1880.

Grieve, M. *A Modern Herbal*. Vol. I and II. New York: Dover Publications, Inc., 1971.

Culinary Herbs and Condiments. New York: Dover Publications, Inc., 1971.

Krochmal, Arnold and Krochman, Connie. *A Guide to the Medicinal Plants of the United States*. New York: Quadrangle/New York Times Book Co., Inc., 1973.

Krochmal, Arnold, Walters, Russel S., and Doughty, Richard M. *A Guide to Medicinal Plants of Appalachia*. Washington D.C.: Government Printing Office, 1971.

Lambert, J., and Muir, T.A. *Practical Chemistry*. London: Heinemann, 1961.

Leopold, Aldo. *A Sand County Almanac*. New York: Ballantine Books, 1970.

Lust, John. *The Herb Book*. New York: Bantam Books, Inc., 1974.

Maurer, David W. *Kentucky Moonshine*. Lexington, Kentucky: The University Press of Kentucky, 1974.

McLeod, Margot. *Handbook of Herbs*. Kingswook, Surrey, U.K.: Elliot Rightway Books, 1976.

Montagne, Prosper. *Larousse Gastronomique*. New York: Crown Publishers, Inc., 1961.

Myers, Philip Van Ness. *Mediaeval and Modern History*. Boston, New York, Chicago, London: Ginn and Company, 1905.

Pederson, Carl S. *Microbiology of Food Fermentations*. Westport, Connecticut: AVI Publishing Co., Inc., 1979.

Porsild, A.E. "Edible Plants of the Arctic," *Arctic*, Vol. VI. (March 1953).

Poucher, William A. *Perfumes and Cosmetics*. New York: D. Van Nostrand Company, 1923.

Ranson, Florence. *British Herbs*. Harmondsworth, Middlesex: Penguin Books Ltd., 1954.

Redgrove, H. Stanley. *Scent and All About It*. London: William Heinemann, Ltd., 1928.

Rimmel, Eugene. *The Book of Perfumes*. London: Chapman and Hall, 1865.

Rose, Anthony H. "The Microbiological Production of Food and Drink," *Scientific American*, Vol. 245, No. 3. (September 1981).

Rosengarten, Frederic, Jr. *The Book of Spices*. New York: Pyramid Books, 1973.

Smith, J. Russell. *Tree Crops, A Permanent Agriculture*. New York: The Devin-Adair Company, 1953.

Solar Energy Research Institute. *National Alcohol Fuels Information Fact Sheets* SIM No. 11015 and 11024. Golden, Colorado: Solar Energy Information Data Bank.

Steele, J. Dorman. *Fourteen Weeks in Chemistry*. New York and Chicago: A.S. Barnes and Company, 1873.

Sturtevant, E. Lewis. *Sturtevant's Edible Plants of the World*. Edited by U.P. Hedrick. New York: Dover Publications, Inc., 1972.

Tannahill, Reay. *Food in History*. Frogmore, St. Albans, Herts (Great Britain): Paladin, 1975.

Thompson, C.J.S. *The Mystery and Lure of Perfume*. London: John Lane The Bodley Head Limited., Detroit: Reissued by Singing Tree Press, Book Tower, 1969.

U.S. Department of Energy, Solar Energy Research Institute. *Fuel From Farms*. Washington, D.C.: GPO, 1980.

U.S. National Alcohol Fuels Commission. *Fuel Alcohol on the Farm*. Washington, D.C.: GPO, 1980.

Verrill, A. Hyatt. *Perfumes and Spices*. Clinton, Massachusetts: L.C. Page and Co., 1940.

Wang, Hwa L., and Hesseltine, C.W. "Wheat Tempeh", *Cereal Chemistry*, Vol. 43, No. 5. (September 1966).

Wang, H.L., Mustakas, G.C., Wolf, W.J., and others. *Soybeans as Human Food–Unprocessed and Simply Processed*. Washington, D.C.: U.S.D.A.

Yonge. *Landmarks of History, Part II, Mediaeval History*. New York: Henry Holt and Company, 1868.

Index

Other Books by Grace Firth

STILLROOM COOKERY
A NATURAL YEAR
LIVING THE NATURAL LIFE

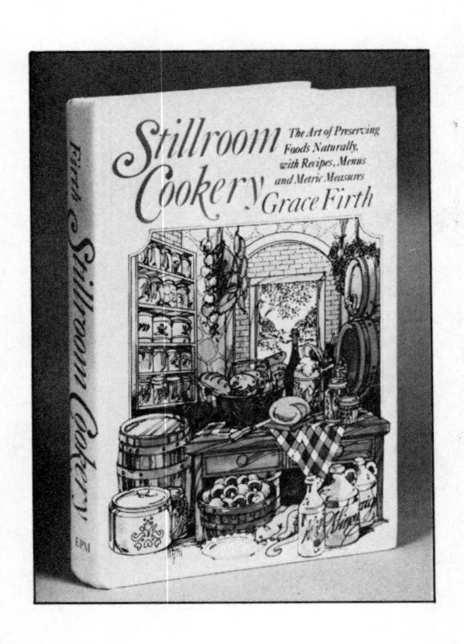

Stillroom Cookery

The companion book to

Secrets of the Still

Before she wrote *Secrets of the Still*, Grace Firth created a major restoration of forgotten tastes and earthy creativity called *Stillroom Cookery*. This is the cookbook, hailed by *Bon Appetit* as "the standard reference work on the subject," that teaches such fascinating things as how to make yeast, clabber milk, bake soda crackers, corn beef, smoke fish, cure sausage, concoct vinegars and more. It describes the techniques for cheesemaking and breadmaking, for jellying and canning, and for drying beans, herbs, peas, teas and nuts.

You can preserve by the natural stillroom way without any special equipment and in no more space than a corner of a basement, a crawl space or the vegetable drawer of a refrigerator. Once you master the stillroom techniques, you move on to Mrs. Firth's menus and recipes (usable with store-bought ingredients if you don't care to put them up yourself). She gives the recipes in metric as well as standard measures so that they can be passed along to upcoming generations. And she serves them all with wry (sometimes outlandish) humor, bits of philosophy and lots of warm-hearted enthusiasm for making the good things of the earth even better.

Stillroom Cookery is a handsome hardcover book with a washable cover. Its price is only $9.95 and it may be ordered direct from the publisher. Send your name and complete address with a check for $11.50, which includes the handling charges, to EPM Publications, Inc., Dept. B, Box 490, McLean, Virginia 22101. Your delight is guaranteed or your money back.